Lecture Notes in Social Networks (LNSN)

For further volumes:
http://www.springer.com/series/8768

Nitin Agarwal • Merlyna Lim • Rolf T. Wigand
Editors

Online Collective Action

Dynamics of the Crowd in Social Media

 Springer

Editors
Nitin Agarwal
Department of Information Science
University of Arkansas at Little Rock
Little Rock
Arkansas, USA

Merlyna Lim
School of Journalism and Communication
Carleton University
Ottawa, Ontario, Canada

Rolf T. Wigand
Departments of Information Science
and Business Information Systems
University of Arkansas at Little Rock
Little Rock
USA

ISSN 2190-5428   ISSN 2190-5436 (electronic)
ISBN 978-3-7091-1339-4   ISBN 978-3-7091-1340-0 (eBook)
DOI 10.1007/978-3-7091-1340-0
Springer Wien Heidelberg New York Dordrecht London

Library of Congress Control Number: 2014946576

Printed on acid-free paper

Springer is part of Springer Science+Business Media (www.springer.com)

Preface

Contemporary collective actions, from the 2011 Egyptian Tahrir uprising to the 2013 Turkish protest in Gezi Park, cannot be separated from the uses of social media. In various recent protests, social media are described as important tools for the disparate crowd to communicate, coordinate, organize, and mobilize itself. How do social media influence, change, or shape (or not) the ways in which people collectively mobilize themselves?

The new phenomenon called "online collective action," a collective action that emerges in a networked online environment, has led to the emergence of a new scholarship. In the last few years, there has been a surge in research devoted to deepening our understanding of online collective action, including our own National Science Foundation-sponsored study entitled "Cyber-collective Movements" (website: http://onlinecollectiveaction.lab.asu.edu/), a project that led to the origin of this book. Analysis of online collective action is a relatively new and fast-moving field due to the rapid development of the technologies in use on the ground. Large gaps in our understanding inevitably remain.

The contemporary use of social media, in particular, has unavoidably linked collective behavior with social networks. As individuals and groups become more networked through social media, real-time communication can take place in the vast social network instantaneously. How does this new way of communication influence and shape the formation of collective action? Is there a need to reassess collective action theory to provide deeper insights for successful contemporary collective action efforts leveraging the new way of communication? Are there any fundamental aspects of contemporary collective action efforts that can be identified and explained using existing collective action theory or do we need innovative approaches and methodologies to reframe and reconceive collective action theory in online environments?

The book aims to contribute to our understanding of contemporary collective action in the age of the Internet and social media by approaching it from various perspectives rooted in diverse disciplines. By carefully selecting a number of qualitative and quantitative studies from computational and social sciences focusing on online collective actions, through this edited volume we aim to promote a symbiotic and synergistic advancement of the multiple, interconnected disciplines. More specifically, we intend to illuminate several fundamental and powerful yet theoretically undeveloped and largely unexplored aspects of collective action in social media.

The book consists of three distinctive parts: (1) concepts, theories, and methodologies; (2) applications; and (3) case studies. The first part consists of high-level discussions on **concepts, theories, and methodologies**. Five chapters have been grouped under this category. The authors of these chapters have proposed, developed, and evaluated the models to advance our understanding of online collective actions beyond anecdotal evidences. The models presented in the chapters have conceptual and theoretical underpinnings from multiple disciplines, including computational sciences, social sciences, mathematics, and statistics to understand, evaluate, and measure online collective actions.

In the chapter entitled "**Sentiment Analysis in Social Media**," Georgios **Paltoglou** discusses methodologies, theories, and techniques from a diverse set of scientific domains, ranging from psychology and sociology to natural language processing and machine learning for sentiment analysis and its contributions to understanding the online collective action phenomenon. The author argues that such actions are motivated by intense emotional states, and sentiment analysis could offer insights to understand the entire life cycle of online collective actions.

Leveraging the notions of natural language processing, **Dipankar Das and Sivaji Bandyopadhyay**, in their chapter entitled "**Emotion Analysis on Social Media: Natural Language Processing Approaches and Applications**," consider collective actions as events, and the emotional change with respect to events is represented using the graphical notion of sentiment-event tracking. The authors mainly focus on blog users and their topics of discussion for identifying emotional changes with respect to time. They have considered the social interactions of the bloggers as collective actions, and their emotional changes based on such actions have been measured from the perspectives of topic and time. **Goldina Ghosh, Soumya Banerjee, and Vasile Palade**, in their chapter entitled "**Discovering Flow of Sentiment and Transient Behavior of Online Social Crowd: An Analysis Through Social Insects**," however, have studied the sentiment flow in a crowd as expressed on social media sites during a crisis event using an agent-based paradigm known as swarm intelligence.

Further, **Anna Chmiel, Julian Sienkiewicz, Georgios Paltoglou, Kevan Buckley, Marcin Skowron, Mike Thelwall, Arvid Kappas, and Janusz A. Holyst**, in their chapter entitled "**Collective Emotions Online**," demonstrate the relevance of negative emotions in sustaining the online discussions in a variety of web platforms, including blogs, discussion forums, IRC channels, and Digg communities. Moreover, the findings suggest individual and collective patterns of emotional activities among web forum users.

The last chapter in this category by **Alan Keller Gomes and Maria da Graça Campos Pimentel**, entitled "**Evaluation of Media-Based Social Interactions: Linking Collective Actions to Media Types, Applications, and Devices in Social Networks**," presents a technique for capturing, representing, and measuring collective actions in social networks. It uses use data mining procedures to measure behavioral contingencies in the form of *if-then* rules and identify links among a variety of platforms for social interactions, such as actions (e.g., like), media objects (e.g., photo), application type (web or mobile), and device type (e.g., Android).

The second part of the book comprises discussions of three **applications**, each extracting and distilling unique angles and perspectives of collective action. These range from a sociological perspective of information sharing resulting in collective knowledge and action to computer-mediated communication in virtual citizen science and to crowdsourcing methods of pro-environmental collective action. All of these application chapters demonstrate how unique outcomes can be achieved using newer information and communication technologies (ICT) coupled with social media to achieve previously unachievable levels and forms of collective action.

The chapter "**The Studies of Blogs and Online Communities: From Information to Knowledge and Action**" by **Emanuela Todeva and Donka Keskinova** is the first among three in this application category. This contribution, while taking a sociological perspective, addresses whether the rise of blogs as a rich information medium may generate new opinion leaders who in turn transform and challenge traditionally held European public views on pharmaceuticals and health care. Todeva's findings reveal that in spite of the relatively high volume of blogs for the investigated period, surprisingly only a small number are interlinked by mutual referrals. The resulting network configuration demonstrates a small core component with many dyads, or short tails, representing a rather fragmented community space. The author concludes that in spite of high technical Internet connectivity, community interaction is relatively limited, and there is no evidence of online crowd or collective action.

The next chapter by **Jason T. Reed, Arfon Smith, Michael Parish, and Angelique Rickhoff**, entitled "**Using Contemporary Collective Action to Understand the Use of Computer-Mediated Communication in Virtual Citizen Science**," looks at how virtual citizen science creates Internet-based projects involving volunteers collaborating with scientists in authentic scientific research. This chapter describes how virtual citizen science can be understood as a form of online collective action. Computer-mediated communication platforms support all aspects of project activity and interaction. Using collective action theory allows for the creation of a collective action space based on a combination of various forms of interaction and project responsibilities available to volunteers.

The third and last chapter in this category of applications authored by **Janis L. Dickinson and Rhiannon L. Crain** is entitled "**Socially Networked Citizen Science and the Crowd-Sourcing of Pro-Environmental Collective Actions**." The authors posit that the social web has changed the nature of human collaboration with new possibilities for massive-scale cooperative endeavors such as scientific research and environmentally important collective action. "Next generation" citizen science practice networks combine crowdsourcing, a joint sense of purpose, and soft institutional governance with the distributed intelligence and efficacy of online social networks. The authors tap into evolutionary theory and social psychology to generate hypotheses for how such "next generation" citizen projects can best support pro-environmental behaviors such as habitat restoration and energy conservation. Finally, the authors explore how properties of social networks themselves enhance the spread of behaviors through the three degrees rule, homophily, social contagion, and the strength of weak ties.

The third part of the book, the **case studies**, provides rich and well-studied illustrations of the ways in which activists and citizen groups in various contexts use online media to collectively organize themselves into political actions. The four chapters included in this category represent four case studies from Spain, Egypt, Morocco, and the United States. The four case studies examine exemplary movement organizations and networks where online tools and practices are woven into the fabric of social movements.

The chapter entitled "**The Spanish 'Indignados' Movement: Time Dynamics, Geographical Distribution, and Recruitment Mechanisms**" presents the first of the four case studies included in this book. Analyzing Twitter data from *the Indignados*, **Javier Borge-Holthoefer, Sandra González-Bailón, Alejandro Rivero, and Yamir Moreno** investigate the mechanisms involved in the emergence, development, and stabilization of the movement. Tracking central issues in the social network literature, such as recruitment patterns and information cascades, this study offers an empirical test to help us answer some sociological questions about collective action and better understand the connection between online networks and social movements.

The second case study continues to deepen our understanding of the usage of Twitter in collective actions. In the chapter entitled "**The Strength of Tweet Ties**," **Rob Schroeder, Sean F. Everton, and Russell Shepherd** explore how Egyptian activists utilized Twitter to frame grievances in ways that resonate with their targeted audience. Using Twitter data collected during the 2011 Egyptian protests, the authors examine whether the Twitter usage helped to diffuse the Arab Spring frame across Egypt and generate greater social cohesion around their messages. Analyzing counts of complete triads, which are regarded as an indicator of cohesion, the authors conclude that activists occupy the position of brokerage in the Twitter network, and this network position has enabled them to frame grievances in ways that brought people together and helped mobilize them for action.

In "**The Arab Spring in North Africa: Still Winter in Morocco?**" **Rebecca S. Robinson and Mary Jane C. Parmentier** present a richly nuanced case study of the use of blogs for political activism in Morocco. Focusing on the Blogoma, the Moroccan Blogosphere, the authors utilize collective action literature to examine the nature of online public narratives in light of several "Arab Spring" protests that occurred in Morocco. They argue that online collective actions through the Blogoma fail to reflect social and political divisions within the society and, thus, are disconnected from the majority of Moroccan citizens. However, this online activism does contribute to a larger and freer civic sphere in the country by providing a space for Moroccan bloggers to exercise their freedom of expression and converse, socially and politically, with each other without an overt control of the government.

The last case study of the book discusses how two groups of American Muslims, the Council of American-Islamic Relations (CAIR) and the Hijabi fashion community, address the issue of hijab and discrimination in the workplace through online collective actions. In "**Online and Offline Advocacy for American Hijabis: Organizational and Organic Tactical Configurations,**" **Rebecca S. Robinson** analyzes the divergent tactics of these two groups by utilizing the concepts of organizational and organic, where the latter is employed to delineate the non- or less hierarchical, self-organized nature of collective action. While neither one can be said to be superior, the author argues that the organic tactics are more grounded in a shared sense of identity and context and are embedded in the experiences of hijabis "on the ground" as opposed to the organizational tactics.

These 12 chapters of the book together provide an interdisciplinary platform for researchers, practitioners, and graduate students from different disciplines to share, exchange, learn, and develop preliminary results, new concepts, ideas, principles, and methodologies, aiming to advance and deepen our understanding of collective action in social media to help critical decision and policy making. The developed methodologies will be a valuable companion and comprehensive reference for anyone interested in newer ICT, examining their role in decision and policy making; understanding the dynamics of interaction, communication, information propagation, and opinion diffusion; and researching in social networks for years to come. The book also serves as an extensive repository of data sets and tools that can be used by researchers leading to a perpetual and synergistic advancement of the discipline.

Recent advancements in Web 2.0 have provided vast opportunities to investigate the dynamics and structure of web-based social networks. Through the publication of this edited volume, we hope to facilitate dissemination of these newly developed methods as well as the investigations and their outcomes.

Little Rock, AR   Nitin Agarwal
Ottawa, ON   Merlyna Lim
Little Rock, AR   Rolf T. Wigand

Glossary

Alaouais	Moroccan royal family.
Behavioral contingencies	Daily interactions between individuals (any type of social interactions) that correspond to observations of what people do or do not do in a variety of situations.
Blogoma	The Moroccan blogosphere.
Collective action frames	The term relates to how collective actors present the issues around which they are collectivizing. This is how collectives discuss why they are organizing and why it is important for their desired objectives to be realized.
Collective action tactics	The term relates to how collectivities share their message, recruit supporters, and are framed by the mainstream. Collectives can take up radical, moderate, violent, peaceful, etc. tactics. They can function principally online or offline. This is what collectives do to advance their causes.
Complex network theory	Complex network theory is the modeling framework that defines a complex system in terms of its subcomponents (nodes) and their interactions (edges).
Complex system	A complex system may be defined as a system whose overall behavior can be characterized as more than the sum of its parts.
Data mining	The extraction and generation of information from the analysis of data.
Dataset	A collection of items (e.g., forum posts, tweets) that have certain properties (e.g., their emotional content has been annotated by a human) and are available for processing by a computer.

Degree distribution	The degrees of all the network's nodes form a degree distribution. In random networks all connections are equally probable, resulting in a Gaussian and symmetrically centered degree distribution. Complex networks generally have non-Gaussian degree distributions, often with a long tail towards high degrees.
Emergence	The manner in which complex phenomena arise from a collection of relatively simple interactions between system components.
Emotion	Emotion is a complex psycho physiological experience of an individual's state of mind as interacting with biochemical (internal) and environmental (external) influences. In humans, emotion fundamentally involves physiological arousal, expressive behaviors, and conscious experience. In psychology and common use, emotion is an aspect of a person's mental state of being, normally based in or tied to the person's internal (physical) and external (social) sensory feeling.
Emotion co-reference	It is a technique to associate the prime components of emotion, i.e., to relate an evaluative expression with its corresponding holder and topic within a text span.
Emotion holder	In linguistics, a grammatical agent or holder is the participant of a situation that carries out the action whereas in computational linguistics, the source of an emotional expression is the speaker or writer or experiencer or the person or organization that expresses the emotions towards a specific topic or event.
Emotion topic	Similar to opinion topic span, emotion topic is also the real-world object, event, or an abstract entity that is the primary subject of the emotion as intended by its holder. Emotion topic also depends on the context in which its associated emotional expression occurs.
Emotion tracking	Emotion tracking means plotting of users' emotions and their intensities on topics or events over time based on the notion of graphical visualization.
Erraji	Mohamed Erraji, Moroccan blogger arrested in 2008 for criticizing the government for corruption and repression of bloggers.
Evaluative expression	Emotions are expressed in natural languages using different language expressions, which are termed as emotional expressions in literature. It is said that sentiment or emotion is typically a localized phenomenon that is more appropriately computed at the paragraph, sentence, or entity level.

February 20th Movement	Moroccan Arab Spring protests, beginning February 20, 2011.
Generic social interactions	Situations that occur frequently in online social network involving users in social interactions. In the representation of these interactions, the actions that users perform in social interactions are detailed.
Hassan II	King of Morocco, 1961–1999.
Hijab	Related to the modest dress and behavior of Muslims. Some Muslims wear headscarves, beards, and loose-fitting, non-revealing clothing because they believe that these are the requirements of modesty set forth by Islam.
ICTs	Information and communication technologies, largely the Internet and cellular telephony.
If-then rule	It is a statement of the form if-then associated with a logical implication that corresponds to a social interaction.
Information cascade	Phenomenon during which individuals in a population exhibit herd-like behavior, deciding to transmit information based on the influence of other individuals (typically neighbors in some network). Although they are generated by different mechanisms, cascades in social and economic systems can be related to avalanche phenomena in physical systems—fracture events, system percolation, etc. In all of them, initial failures increase the likelihood of subsequent failures, leading to eventual outcomes that are inherently difficult to predict, even when the properties of the individual components are well understood. A cascade can be modeled as an information propagation graph.
Islamophobia	Modern-day manifestations of Orientalism, related to irrational fear of Islam and Muslims, often resulting in online hate speech and offline discrimination and violence. In multicultural societies in which racism and sexism have become illegitimate forms of discrimination, Islamophobia is a form of acceptable discrimination because, while cisgender and race are not perceived as choices, religion is.
Machine learning	A subfield of Artificial Intelligence, concerned with getting computers to learn to do things from data, without being explicitly programmed by a human.
Makhzen	Traditional name for the Moroccan government institution, which includes the monarchy and the Moroccan elite.

Media-based social interactions	Interactions between users of online social networks that occur from sharing different types of media.
Message graph	Every message transmitted in a social media has a unique identity and relevance. The individual message generated and sent by a member can be represented as an object in the graph. Message graph assists in analysis of node along with the transmission of information, giving a clear image of the expected temporal tendency of the social crowd.
Modularity	Many complex networks consist of a number of modules. There are various algorithms that estimate the modularity of a network, many of which are based on hierarchical clustering. Each module contains several densely interconnected nodes, and there are relatively few connections between nodes in different modules.
Mohamed VI	King of Morocco, 1999–present.
Mourtada	Fouad Mourtada, Moroccan engineer arrested in 2008 for impersonating the King's brother on Facebook.
Natural Language Processing (NLP)	A subfield of computer science and linguistics aimed at analyzing and understanding language.
Network	A network is a physical system that can be mathematically represented by a graph comprising a set of N nodes and E edges.
Node degree	The degree of a node is the number of connections that link it to the rest of the network. This is the most fundamental network measure, and most other measures are ultimately linked to node degree.
Opinion mining	The computational treatment of subjectivity (e.g., opinions, emotions, evaluations) in text.
Organic solidarity	The term originated with Emile Durkheim in regard to non-institutional forms of solidarity among socioeconomic classes. The term can also be applied to other non-hierarchic, loosely defined groups that collectivize around various topics.
Pheromone communication	Ant and swarms demonstrate well-defined chemical communication signal known as *pheromone* to segregate and distinguish specific communication pattern from cells of high concentration to those of low concentration. Hence, the *positive and negative sentiment* of transient *crowd* could be modeled and the local influence can be measured on their posts through pheromone modeling and reinforcement of the shortest path of an ant or swarm's life cycle.
PJD	Party of Justice and Development, legal Islamist party in Morocco.

Scale-free network	Graph with a power-law degree distribution, i.e., a statistical distribution of the form $p(x) \sim x^{-y}$. "Scale-free" means that degrees are not grouped around one characteristic average degree (scale) but can spread over a very wide range of values, often spanning several orders of magnitude.
Sentiment and opinion flow	In the current world of social networking, the meaning of communication or interaction denotes the sharing of sentiment (like happy, sad) and placing opinion on a particular event irrespective of geographical boundary, language, and age. Sentiment analysis and opinion flow can also be applied on offline events.
Social crowd	Rapid sharing of views in social media signifies the sentiment and emotional status of the crowd on a specific scenario. The crowd participates in the interaction which gradually grows "on fly", and once the event fades out the crowd slowly disappears from the social media.
Social interactions	Actions or practices of two or more people mutually oriented, i.e., they are behaviors that affect people's subjective experiences or intentions of other people. In the scope of our work, social interactions are behavioral contingencies represented as logical implications that can be described as if-then rules.
Social media	It refers to interaction among people in which they create, share, and/or exchange information and ideas in virtual communities and networks. It is a group of Internet-based applications that build on the ideological and technological foundations of Web 2.0 and that allow the creation and exchange of user-generated content.
Social stimulus	It is the action that initiates a social interaction.
Transient behavior	The majority of the population in any social network can get involved in any momentary event where they share and provide opinion or intimate behavior. This kind of behavior is temporary and recedes as soon as its importance diminishes.
Valence	In psychology, especially in discussing emotions, the term refers to the intrinsic attractiveness (positive valence) or aversiveness (negative valence) of an event, object, or situation.

Acknowledgments

This book would not have been completed without the enthusiastic support of many people and institutions. We express our deep gratitude to the esteemed authors for contributing chapters representing an extensive range of concepts, theories, tools, methodologies, applications, and case studies from multi- and inter-disciplinary perspectives of collective action. Considering the inter-disciplinary nature of the contributions, the book has significantly benefited from the outstanding contribution of the reviewers, to whom we extend our special thanks.

We thank Reda Al-Hajj and Nasrullah Memon, editors for the Springer Lecture Notes in Social Network (LNSN), for their enthusiastic support in including this publication in the LNSN series. We truly appreciate the timely help and assistance provided by Springer staffs in the Netherlands and Austria, especially Annelies Kersbergen, Stephen Soehnlen, and Levente Istvan Koltai, as well as the support of numerous graduate students at Arizona State University and the University of Arkansas at Little Rock.

We would like to acknowledge the financial support for this book project provided by the United States National Science Foundation's Social-Computational Systems program and Human-Centered Computing program (award numbers IIS-1110868 and IIS-1110649).

Lastly, we are grateful to the support of our respective intellectual "homes." Nitin Agarwal and Rolf Wigand would like to thank the Department of Information Science and the Jerry L. Maulden-Entergy Fund at the University of Arkansas at Little Rock. Merlyna Lim would like to express her gratitude to the School of Journalism and Communication at Carleton University and the Consortium for Science, Policy and Outcomes and the School of Social Transformation at Arizona State University. Part of this book project was completed while Merlyna was a visiting scholar at the Center for Information Technology at Princeton University. She would like to extend her personal appreciation to the Center, which generously hosted her in the academic year 2013–2014.

Contents

List of Contributors

Sivaji Bandyopadhyay Jadavpur University, Kolkata, India

Soumya Banerjee Birla Institute of Technology, Ranchi, India

Javier Borge-Holthoefer Universidad de Zaragoza, Zaragoza, Spain

Kevan Buckley University of Wolverhampton, West Midlands, UK

Anna Chmiel Warsaw University of Technology, Warszawa, Poland

Rhiannon L. Crain Cornell University, Ithaca, NY, USA

Maria da Graça Campos Pimentel University of São Paulo, São Paulo, Brazil

Dipankar Das National Institute of Technology, Shillong, India

Janis L. Dickinson Cornell University, Ithaca, NY, USA

Sean F. Everton Naval Postgraduate School, Monterey, CA, USA

Goldina Ghosh Birla Institute of Technology, Ranchi, India

Alan Keller Gomes University of São Paulo, São Paulo, Brazil

Sandra González-Bailón Oxford University, Oxford, UK

Janusz A. Hołyst Warsaw University of Technology, Warszawa, Poland

Arvid Kappas Jacobs University Bremen, Bremen, Germany

Donka Keskinova Plovdiv University, Plovdiv, Bulgaria

Yamir Moreno Universidad de Zaragoza, Zaragoza, Spain

Vasile Palade University of Oxford, Oxford, UK

Georgios Paltoglou University of Wolverhampton, West Midlands, UK

Michael Parish Adler Planetarium, Chicago, IL, USA

Mary Jane C. Parmentier Arizona State University, Phoenix, AZ, USA

Jason T. Reed Adler Planetarium, Chicago, IL, USA

Angelique Rickhoff Adler Planetarium, Chicago, IL, USA

Alejandro Rivero Universidad de Zaragoza, Zaragoza, Spain

Rebecca S. Robinson Arizona State University, Phoenix, AZ, USA

Rob Schroeder Naval Postgraduate School, Monterey, CA, USA

Russell Shepherd Naval Postgraduate School, Monterey, CA, USA

Julian Sienkiewicz Warsaw University of Technology, Warszawa, Poland

Marcin Skowron Austrian Research Institute for Artificial Intelligence, Vienna, Austria

Arfon Smith Adler Planetarium, Chicago, IL, USA

Mike Thelwall University of Wolverhampton, West Midlands, UK

Emanuela Todeva University of Surrey, Surrey, UK

Part I
Concepts, Theories, and Methodologies

Sentiment Analysis in Social Media

Georgios Paltoglou

Abstract Sentiment Analysis deals with the detection and analysis of affective content in written text. It utilizes methodologies, theories, and techniques from a diverse set of scientific domains, ranging from psychology and sociology to natural language processing and machine learning. In this chapter, we discuss the contributions of the field in social media analysis with a particular focus in online collective actions; as these actions are typically motivated and driven by intense emotional states (e.g., anger), sentiment analysis can provide unique insights into the inner workings of such phenomena throughout their life cycle. We also present the state of the art in the field and describe some of its contributions into understanding online collective behavior. Lastly, we discuss significant real-world datasets that have been successfully utilized in research and are available for scientific purposes and also present a diverse set of available tools for conducting sentiment analysis.

1 Introduction

The unprecedented rise in the popularity of social media platforms (Harvey 2010), exemplified by blogs, forums, and services such as Facebook, Twitter, Google Plus, etc. and their infiltration in everyday life (e.g., Fouche 2011; Barnett 2009) has resulted in an important paradigm shift in the way that people communicate with each other online and, more generally, interact with the Web. Previously, users were typically limited to consuming content authored by professionals, such as news agencies or corporations. In contrast, nowadays they can effortlessly create

G. Paltoglou (✉)
School of Technology, University of Wolverhampton, Wulfruna Street, Wolverhampton WV1 1LY, UK
e-mail: g.paltoglou@wlv.ac.uk

N. Agarwal et al. (eds.), *Online Collective Action*, Lecture Notes in Social Networks, DOI 10.1007/978-3-7091-1340-0_1, © Springer-Verlag Wien 2014

3

and share their own content and seamlessly interact with other users within a network of peers, often in a real-time, synchronous manner.

The importance of this phenomenon and its repercussions to society were vividly demonstrated in 2011, when a series of sociopolitical events, such as the London riots[1] and the Arab Spring,[2] took place. In both cases, online social media were regarded to significantly contribute to the emergence and proliferation of the events, with one participant of the latter event claiming that "We use Facebook to schedule the protests, Twitter to coordinate and YouTube to tell the world" (Howard 2011). Those effects were more pronounced by the fact that even official authorities considered (Halliday and Garside 2011) or took direct action (Gazzar et al. 2011) in shutting down Internet communications in order to prevent people from having access to such services.

In this chapter, we analyze the effects and implications that the new field of sentiment analysis can have in this novel environment. The next section provides a concise but thorough introduction to the field, and in Sect. 3 we discuss the applications of the field in social media in general and in the context of online collective behavior, more specifically. Sections 4 and 5 present some important, real-world datasets and tools that have been successfully used by researchers and are freely available, and in Sect. 6 we conclude and summarize.

2 Sentiment Analysis

Sentiment analysis is a subdiscipline within data mining, machine learning, and computational linguistics and also borrows elements from psychology and sociology. Generally, it deals with the computational treatment of expressions of opinion, sentiment, emotion, beliefs, and speculation in written text (Wiebe et al. 1999). Those are concisely defined as *private states*, i.e., states that are not open to objective observation or verification (Quirk et al. 1985). The field has also been referenced in research as *opinion mining* or *subjectivity analysis*,[3] and we will use those terms interchangeably in this chapter.

More specifically, opinion mining addresses the problem of detecting, extracting, analyzing, and quantifying expressions of private states in written text in an automatic, computer-mediated fashion. Particular emphasis should be placed on the term "computer-mediated," as the field has a particular focus on designing, analyzing, and implementing software that performs the aforementioned analysis in an automatic manner.

---

[1] http://en.wikipedia.org/wiki/2011_England_riots

[2] http://en.wikipedia.org/wiki/Arab_Spring

[3] For more information about the terminology in the field, we refer the interested reader to chapter 1.5 of Pang and Lee (2008).

As a result, sentiment analysis software receives as input unstructured data, such as the textual exchanges between users (e.g., tweets, Facebook updates, blog/forum posts, etc.) which by themselves are of use only to language specialists, and provides as output an informed estimate of the sentiment contained within the exchange. That estimate can take a diverse set of forms depending on a number of factors, such as the specific prerequisites of the application, the domain of interest, the psychological paradigm adopted, etc. Typical examples of valued output can include but are not limited to:

- A *binary* decision indicating whether the affective content of the communication belongs to one of two predefined categories. For example, an opinion about a new political legislation can have a positive or negative position about it, supporting or rejecting it, respectively (Whitelaw et al. 2005). In some applications, such as online discussions, where not all exchanges are necessarily affective (e.g., Chmiel et al. 2011a; Mitrović et al. 2011), a *ternary* scheme is more appropriate and adopted: {objective, positive, negative}, where the *objective* category typically signifies the absence of opinionated or affective content, such as encyclopedic-type, mainly informative content.
- A *real* value providing more fine-grained and detailed information about affective content. Typical applications can include studies on the level of *valence* or *arousal* at a specific scale (e.g., [1,9]) expressed in a forum post (Gonzalez-Bailon et al. 2010; Dodds and Danforth 2009; Paltoglou et al. 2013). Valence is defined as the dimension of experience that refers to hedonism (i.e., pleasure and displeasure) and arousal refers to the level of excitement or energy of the individual (Barrett and Russell 1999).
- A *categorical* classification where the analysis aims to determine the general psychological state of the author of a message. Typically, the analysis will involve several potential states such as nervousness, anxiety, fear, fatigue, and tension (Mishne 2005; Bollen et al. 2011). In the same manner, *basic emotions*, such as love, hate, etc. (Dalgleish and Power 1999) can be detected in written text (Strapparava and Mihalcea 2008), although there is significant debate within the field of psychology on the human agreement (Strapparava and Mihalcea 2007) and universality (Mauss and Robinson 2009) of such states.

Sentiment analysis is a nontrivial task, as even people often disagree on the affective content of written text (Paltoglou et al. 2010; Strapparava and Mihalcea 2007). Prosaic elements, such as irony and thwarted expectations (occurring when a change of opinion takes place in the end) pose particular challenges. Contextuality is also often vital; a review comprising only of the sentence "go read the book!" would be positive in a book review, but negative if referring to a movie. People also often find unique ways of expressing affect without necessarily using affective words and occasionally communicate ambiguous messages.

There are also additional issues pertaining to social media, because their typical content does not necessarily conform to the standard syntactic and grammar rules. In contrast, it contains idiomatic expressions which varies significantly based on the users' social background (Thelwall 2009), heavily utilizes acronyms and

Table 1 Examples of textual communication with affective content

Source	Text
MySpace	hey witch wat cha been up too
Twitter	It saddens my heart 2 C women with black eyes! :(
YouTube	were the f... are the lyrics u suck man...but the song is great..SOAD forever
Forums (BBC)	And I think you are a slanderous and ignorant fool... Great how free speech works, isn't it? ... I love it!
Digg	This story makes me happy and sad at the same time...

emoticons, and is overall highly heterogeneous and often targeted to specific social groups. Table 1 presents some examples of the aforementioned challenges from a variety of social media.

2.1 Behind the Scenes

Machine Learning techniques (Chen and Zimbra 2010; Sebastiani 2002) have been an integral part of opinion mining, as a significant number of sentiment analysis solutions are based on them. According to this approach, a general inductive process initially *learns* and stores the characteristics of a category (e.g., opinions in favor of some legislation) during a training phase. This is achieved by observing the properties of a set of humanly annotated, preclassified text segments. Those preclassified text segments, which can be forum posts, political speeches, etc., comprise the *training dataset.*

Creating such a dataset is generally a time-consuming task, as it typically requires manual, human effort in order for the text segments to be read, understood, and assigned to a category. Nonetheless, there are ways in which the process can be done in an automatic or semiautomatic way, for example, by examining the metadata that accompany the textual message, such as the "number of stars" in product reviews (Pang et al. 2002) or the ideological stand or final vote in political issues (Thomas et al. 2006). Alternatively, implicit signals within the message itself, e.g., the type of emoticons used (Pak and Paroubek 2010) can be used to infer an overall affective state. Lastly, crowd sourcing techniques can provide an alternative solution to producing such annotations (Brew et al. 2010).

The knowledge that is acquired through the training phase is later *applied* to determine the best category for new, unseen text segments (Sebastiani 2002). Based on this general theoretical foundation, a number of sentiment analysis techniques have been presented that utilize specific machine learning algorithms, such as Naive Bayes (John and Langley 1995), Logistic Regression (Le Cessie and Van Houwelingen 1992), Support Vector Machines (Platt 1999; Joachims 1999), and others. A detailed discussion on machine learning is beyond the scope of this book, but we refer the interested reader to the books of Mitchell (1997) and Bishop (2006) for a thorough introduction to the topic.

Often, sentiment analysis approaches that utilize machine learning take advantage of the idiosyncrasies of affective communication and extend the standard features/properties of analyzed documents with additional, sentiment-based elements (Velikovich et al. 2010; Whitelaw et al. 2005; Agarwal et al. 2011). For example, limited human assistance can be employed in annotating specific emotionally definitive phrases (Zaidan et al. 2007) or analyzing the syntax of the text in order to extract useful patterns (Wilson et al. 2005b). More often than not, such properties can be extracted from *emotional* dictionaries; lexicons in which lemmas are annotated with affective semantics, for example, the level of positivity or negativity they typically convey.

There is a significant number of such lexicons in research, which have been produced either automatically or semiautomatically (Jijkoun et al. 2010), that usually extend the WordNet (Miller 1995)[4] lexical database with additional annotations. Examples include WordNet-Affect (Strapparava and Valitutti 2004) and SentiWordNet (Baccianella et al. 2010) which adopt a different annotation scheme. The former contains 4,787 words, mainly nouns and verbs, that directly or indirectly refer to mental states. For example, the term "anger" is annotated as referring to "emotion," while "cry" belongs to the "behavior" category. SentiWordNet, on the other hand, focuses on a simpler ternary scheme and gives each lemma three scores based on how positive, negative, or objective it is. The three scores sum up to 1, giving the annotations an interesting probabilistic interpretation. For example, the noun "love" has a positive value of 0.625 and negative value of 0.0, while "hate" has a negative value of 0.75 and a positive value of 0.0.

In addition, there are also affective lexicons that were produced by psychological studies and are manually annotated by human coders. Those include the "Linguistic Inquiry and Word Count" or LIWC (Pennebaker and Francis 1999) and the "Affective Norms for English words" or ANEW (Bradley and Lang 1999) lexicons which also offer different types of annotations. LIWC classifies words in one or several, not necessarily affective, categories, such as social, family, time, positive, anger, etc. ANEW provides for each word three values of valence, arousal, and dominance on a [1–9] scale. Both have been used in a number of large-scale studies (e.g., Owsley et al. 2006; Bollen et al. 2011; Gonzalez-Bailon et al. 2010).

Affective lexicons have also been utilized in nonmachine learning solutions to opinion mining. Typically, such approaches can be very effective in cases where training data is particularly scarce. In addition, the fact that they do not need training and can often be applied *off the shelf* is often seen as a significant advantage. They have been known to perform adequately effectively in a number of diverse environments (Paltoglou and Thelwall 2012), often reaching human-level accuracy in certain cases (Thelwall et al. 2010). Examples include the Opinion Observer (Ding et al. 2008), which is mostly aimed at estimating the polarity of product reviews, the Semantic Orientation CALculator or SO-CAL software

---

[4] WordNet is a lexical database of English words which in addition to standard definitions also provides semantic relations between words.

(Taboada et al. 2001), that provides a ternary scheme, and SentiStrength (Thelwall et al. 2010), which estimates the strength of positive and negative sentiment by producing two separate values: -1 (not negative) to -5 (extremely negative) and 1 (not positive) to 5 (extremely positive). SentiStrength was designed for application in social media exchanges, where as noted before, communication is typically short and informal.

Most lexicon-based approaches, as those discussed above, are usually based on estimating the sentiment content of textual communication by utilizing one or more affective lexicons. In addition, in order to provide increased effectiveness, they also incorporate syntactical-based capabilities, such as detecting negation and emoticons and/or typically incorporate lists of intensifier/diminisher words that increase or decrease respectively the affective strength of sentiment words. Such additional capabilities would provide them the ability to distinguish the difference of valence between phrases such as "I don't love you!" and "I love you very much!", even though both contain the same affective word "love."

3 Sentiment Analysis in Social Media

The aforementioned increase of user-generated content has resulted in creating a digital landscape where the application of opinion mining can provide invaluable information and insight about the affective state of individuals. As online participation nowadays very often accompanies offline activity, the results can provide useful insights into their general well-being and behavior (Kramer 2010). Importantly, applied in massive scale (e.g., Godbole et al. 2007; Kramer 2010), sentiment analysis can provide useful insights about groups or collectives of people.

The means through which it can produce such analyses is by processing the textual content of online social media communication and providing concrete and quantifiable information about its affective properties. Group-level analysis can be obtained by automatically analyzing the *public* communication between individual group members or their messages to the outside world and aggregating the results in order to formulate an overview of the collective affective properties of the group's communication.

The produced analysis can be of different granularity levels, depending on the specific requirements of the application and the interests of the analyzing party. For example, individuals may be grouped together by the emotional properties of their communication (Chmiel et al. 2011a), their political stand (Thomas et al. 2006), or by the discussion threads they participate in (Gonzalez-Bailon et al. 2010). Typically, it is useful to trace the textual exchanges of group members through time, effectively creating a temporal-based analysis in regard to affective communication in order to observe the progression of emotion throughout the life cycle of important events (Diakopoulos and Shamma 2010; Bautin et al. 2008; Chang et al. 2011).

As most collective actions are strongly linked to, typically negative, affective states, such as anger, indignation, or high levels of arousal (Russell 1980), such an analysis can provide unique insights throughout the formulation, development, and death of collective phenomena. This information can be of paramount importance in understanding the intricate workings of collective action by providing evidence of specific affective states (e.g., negativity) during the life span of such actions. For example, an analysis of tweets relating to the 2011 London riots showed that Twitter was mostly used to notify users about subsequent events, rather than promote illegal activity (Tonkin et al. 2012).

Chmiel et al. (2011a) study the role of emotions in the life span of online communities. They investigate whether the affective properties of the communication amongst members can quantitatively and qualitatively influence the community's trajectory in time (i.e., whether they will dissolve quickly or persist). They cluster posts from blogs, forums, and other social media based on the similarity of their emotional valence, and one of the findings they present is that the length of such clusters can be significantly affected by their emotional properties, compared to a random clustering. Based on those results, they conclude that collective emotional states created and propagated in online communities can be of vital importance for the survival of those communities. In a similar fashion, Mitrović et al. (2011) show that strong, negative emotions play a critical part in the formulation, survival, and growth of online communities. Specifically, they demonstrate that one of the driving mechanisms of thriving communities is the, mostly negative, emotional state and communication of their users, a finding that is also confirmed by Gonzalez-Bailon and Paltoglou (2012). They, in turn, study the whole life cycle of an online community from birth to dissipation and show that negative posts and their authors tend to be more popular than positive ones. They also conduct a longitudinal analysis to show that increases of positive comments have a negative impact on the activity of the community, concluding that, potentially in contrast to popular belief, disagreement and discontent are crucial to keep communities together and alive.

Twitter, as expected, has attracted significant attention in regard to the affective properties of the exchanges of its users (Agarwal et al. 2011). Bollen et al. (2011) view tweets as "temporally-authentic microscopic instantiations of public mood state" and attempt to extract six dimensions of mood from them: tension, depression, anger, vigor, fatigue, and confusion. They discover that public mood is indeed closely correlated with wider social and economical phenomena, such as stock market and crude oil prices. Importantly, they also report that significant events in the social or political arena have direct and measurable effects in the public mood as expressed in tweets. Thelwall et al. (2011) apply SentiStrength (Thelwall et al. 2010) in Twitter and discover that popular events, regardless of their actual nature, are typically associated with increases in negative sentiment strength, confirming the aforementioned findings on the importance of negativity in online communities. Diakopoulos and Shamma (2010) develop an analytical methodology for understanding the temporal dynamics of sentiment in relation to televised events (in the particular case, presidential debates) and demonstrate visuals and metrics that can be utilized to measure aggregated group emotions, anomalies, and indications of controversial topics. On the same topic, Pennebaker (2008) and Pennebaker and Persaud (2010) are able to provide useful insights into the psychology of the candidates by analyzing their words and the frequency with which they are used during the discussions.

Kramer (2010) examines the use of emotion-barring words for 100 million Facebook users and reports that their usage closely follows self-reported satisfaction with life and that expected sentiment peaks occur at emotionally significant celebrations (e.g., Thanksgiving for U.S. users). Gonzalez-Bailon et al. (2010) show that automatic estimations of sentiment of public opinion can predict presidential approval rates, indicating a strong connection between online sentiment and official polls. More generally, they show that political events can incite changes in emotional perceptions and that these perceptions can shape political attitudes. Similarly, O'Connor et al. (2010) show that sentiment analysis techniques can predict political opinion and consumer confidence with high correlation, showing that they provide a good estimator of public feeling.

Mishne and de Rijke (2006) capture the "blogosphere state-of-mind" by analyzing and aggregating the affective content of 3.5 million blog posts over 39 days from data extracted from LiveJournal,[5] one of the largest blogging communities. They provide two case studies, one from the 7 July 2005 London bombing and a second from a recurring event, in the particular case examining the drinking habits of bloggers during the weekend. Their analysis shows that in the former case there is a significant increase in the reporting of negative emotions, such as "sadness" and "shock" while at the same time there was a significant decrease of positive emotions. They were also able to detect the *irregular* mood behavior that resulted from the event. In the latter case, of a recurring phenomenon, they were able to detect spikes of frequency of words relating to increased alcohol consumption during the weekend periods. Overall, their results strongly indicated that changes in public mood in relation to either irregular events or recurring events can be detected in the online communication of social media users.

In conclusion, it can be seen that the application of sentiment analysis in social media communication can provide invaluable information concerning the emotional state of individuals and collectives. The discussed research has shown that emotions play a vital role in the survival of online communities and often accompany significant worldwide sociopolitical events. Through the application of opinion mining, such events can be effectively detected and their perception by the general public extracted and quantified.

---

[5] http://www.livejournal.com

4 Datasets

In this section, we introduce and briefly discuss some of the datasets that have been used in the past in research. When the discussed dataset is publicly available, we'll make a note of it.

Ideally, researchers would like to have direct access to social network sites in order to be able to conduct research with all available data. For various reasons, such as user privacy or corporate policies, that is rarely possible, especially when the site is closed, i.e., users can only access the data of limited other users that have explicitly allowed them to do so, e.g., Facebook. There are studies nonetheless (Kramer 2010; Velikovich et al. 2010) that are typically conducted by employees of such services that often provide significant insights because of the massive scale of the analysis that they are based on.

Twitter on the other hand provides significant advantages over Facebook, because everyone can access the majority of its contents, since most of it is public. In fact, as we've seen, content from the site has been used numerous times by researchers (Bollen et al. 2010, 2011; Pak and Paroubek 2010; Paltoglou and Thelwall 2012). Unfortunately, its Terms of Service[6] prohibit the distribution of any content to third parties; therefore, any collected data is usually prevented from being given to other researchers. Thankfully, the Twitter API[7] provides easy access to the site's content through programming. In a similar fashion, the YouTube API[8] also allows programs to perform most of the operations available on the actual site, such as searching for videos, retrieving feeds, seeing relevant content, etc. Therefore, even though it is not always easy to distribute data from social media, for a significant number of sites there are ways in which such data can be collected and analyzed by individual researchers.

Despite the aforementioned limitations, there is a significant number of publicly available datasets for research purposes. Paltoglou et al. (2010) present two datasets extracted from BBC discussion forums and the Digg Web site. The former includes about 2.5 million textual exchanges on 100 K discussion threads, such as politics, religion, and UK and World news, spanning four years, from June 2005, when the forum went online, until June 2009, when part of it was shut down. Digg is a social news Web site, where people link to news articles they find interesting, and other users discuss them on the Web site or vote them up or down. The specific dataset comprises a full three month crawl of all activity on the Web site, from February to April 2009 and includes about 1.2 million stories and 1.6 million individual comments. Both datasets have been extensively used in research in studies of online collective phenomena (Chmiel et al. 2011a, b; Mitrović et al. 2011). More details on

---

[6] https://twitter.com/tos

[7] API stands for "Application Protocol Interface" and usually provides methods for accessing the content of services or software through programming techniques. A guide to the Twitter API can be found here: https://dev.twitter.com/

[8] http://code.google.com/apis/youtube/overview.html

the datasets as well as information on how they can be obtained can be found in Paltoglou et al. (2010).

Another set of important datasets are the ICWSM Spinn3r Datasets (Burton et al. 2009). There are two versions of the datasets, one from 2009[9] and a more recent one from 2011.[10] Both datasets are provided by Spinn3r.com and include several million blog posts crawled by Spinn3r. The former dataset comprises of 44 million blog posts, published between 1 August and 1 October 2008, to an uncompressed size of 142 GB. It is offered in a preprocessed XML format, in order to facilitate researchers analyze its content. Its time span includes a number of significant worldwide news events, such as the Olympic Games, the U.S. presidential nominating conventions, the beginning of the credit crisis, etc. The 2011 dataset (Burton and Soboroff 2011) comprises of 386 million blog posts, new articles, forum posts, and social media communication published between 13 January and 14 February 2011, to an uncompressed size of approx. 3 TB. Its time span includes a significant number of events relating to the Arab Spring, including the Egyptians protests, the Tunisian revolution, and others. Both datasets have already been used in sentiment analysis studies and other types of research (e.g., Cha et al. 2009) and are available for research purposes.

5 Sentiment Analysis Tools

There are a significant number of machine learning tools that are freely available for research, and sometimes industrial, purposes. Most of those either provide optimized versions of specific classification algorithms or, alternatively, general frameworks incorporating several classifiers.

Typical examples of optimized classifiers include SVM^{light} (Vapnik 1999; Joachims 1999), LIBSVM (CC. Chang and Lin 2011), and LIBLINEAR (Fan et al. 2008) all of which provide different implementations of Support Vector Machine (SVM) classifiers, which are considered state of the art in terms of classification accuracy. LIBLINEAR in particular is explicitly designed for application in massive datasets with millions of documents and features, making it an ideal candidate for such environments.

General frameworks of classifiers include Weka (Hall et al. 2009) which provides an extended set of machine learning algorithms (SVM, Rule-based algorithms, etc.) for data mining tasks. In addition to classification, Weka provides tools for clustering, regression, and visualization, making it a very capable, all-around tool for analyzing data. Other tools that have been used in research and are freely available include LingPipe (Alias-i 2008), Mallet (McCallum 2002), and Apache

---

[9] Available at: http://www.icwsm.org/2009/data/index.shtml

[10] Available at: http://icwsm.org/data/index.php

OpenNLP[11] all of which are fully fledged toolkits for processing text using natural language processing and machine learning techniques. They both provide a wealth of tools for processing text, such as named-entity recognition, part-of-speech tagging, sentence segmentation, etc. (Manning and Schütze 1999).

As most of these tools implement machine learning techniques, they will typically require some sort of training in order to be applied in realistic scenarios, therefore, a *training dataset* is necessary (see Sect. 2.1). In rare cases, already trained models for certain tasks are provided with some of the tools. For example, LingPipe provides trained models for part-of-speech tagging and Chinese word segmentation,[12] but to our knowledge no trained models for sentiment analysis are available.

Lexicon-based approaches provide a potential solution to the issue, since in most cases they can by applied *off the shelf* without any training or need for human labor. Typical examples include the lexicon-based classifier by Paltoglou and Thelwall (2012), SentiStrength (Thelwall et al. 2010), and OpinionFinder (Wilson et al. 2005a). The former two[13] have been explicitly designed and tested for the type of informal textual communication that is typical in online discussions, tweets, and social network communication. The former provides a ternary classification scheme where a textual segment is classified as *{objective, positive, negative}* and the latter is mainly focused on measuring the level of positivity and negativity in text. The OpinionFinder[14] system performs various levels of affective analysis, by identifying subjective sentences and extracting various aspects in relation to them, such as the opinion holder (i.e., the entity expressing an opinion), the opinion target, etc.

6   Summary and Conclusions

In this chapter, we focused on the field of sentiment analysis and demonstrated how it can contribute to deepening our understanding of online communities and collective actions. We discussed how the research can be utilized to analyze such phenomena throughout their life cycle and provided insights into the internal mechanisms that drive them. We also presented some of the results that have already been produced, demonstrating the importance of emotions in the creation, dissemination, survival, and dissipation of online communities and collective actions. It is interesting to note that those results have come from a variety of scientific domains, e.g., sociology, complex systems and physics, indicating the

---

[11] http://incubator.apache.org/opennlp/

[12] http://alias-i.com/lingpipe/web/models.html

[13] Available at: http://www.cyberemotions.eu/data.html

[14] Available at: http://code.google.com/p/opinionfinder/

diversity of the analyses that is possible on the output of opinion mining algorithms applied to social media communication.

Additionally, we presented relevant, real-world, large-scale datasets that are publicly available for interested researchers. Some of them are focused, containing information and content from specific social media Web sites for varied periods of time, which extend from some months to several years, while others are more general and are comprised of data extracted from a wealth of sources, such as blogs, forums, and discussion boards, which extend a limited amount of time. We also discussed how despite the fact that some previously used datasets are not available for distribution due to the Terms of Services of individual social media Web sites, programming interfaces are typically provided that can give access to their contents.

Lastly, we presented the state-of-the-art in efficient and effective ways of conducting sentiment analysis in social media, using either machine-learning techniques or unsupervised, lexicon-based approaches. Importantly, we presented freely available tools that can be used for sentiment analysis tasks, either after some training, in the case of the former approaches, or *off the shelf* for the latter solutions.

References

Agarwal A, Xie B, Vovsha I, Rambow O, Passonneau R (2011) Sentiment analysis of Twitter data. In: Proceedings of LSM'11. Association for Computational Linguistics, Stroudsburg, PA, pp 30–38

Alias-i (2008) Lingpipe 4.1.0. http://alias-i.com/lingpipe

Baccianella S, Esuli A, Sebastiani F (2010) Sentiwordnet 3.0. In: Proceedings of LREC'10, Valletta, Malta

Barnett E (2009) Facebook fuelling divorce, research claims (21 Dec, 2009, The Telegraph). http://www.telegraph.co.uk/technology/facebook/6857918/Facebook-fuelling-divorce-research-claims. html. Accessed 08 Sept 2011

Barrett LF, Russell JA (1999) The structure of current affect: controversies and emerging consensus. Curr Dir Psychol Sci 8(1)

Bautin M, Vijayarenu L, Skiena S (2008) International sentiment analysis for news and blogs. In: Proceedings of ICWSM'08, Seattle, Washington, DC

Bishop CM (2006) Pattern recognition and machine learning (information science and statistics). Springer, New York

Bollen J, Mao H, Zeng XJ (2010) Twitter mood predicts the stock market. CoRR. http://arxiv.org/abs/1010.3003

Bollen J, Mao H, Pepe A (2011) Modeling public mood and emotion: Twitter sentiment and socio-economic phenomena. In ICWSM, Barcelona, Spain

Bradley MM, Lang PJ (1999) Affective norms for English words (ANEW): instruction manual and affective ratings. The Center for Research in Psychophysiology, Gainesville, FL

Brew A, Greene D, Cunningham P (2010) Using crowdsourcing and active learning to track sentiment in online media. In: Proceedings of ECAI'10, Lisbon, Portugal, pp 145–150

Burton KNK, Soboroff I (2011) The ICWSM 2011 spinn3r dataset. ICWSM, Barcelona

Burton K, Java A, Soboroff I (2009) The ICWSM 2009 spinn3r dataset. ICWSM, San Jose, CA

Cha M, Pérez JAN, Haddadi H (2009) Flash floods and ripples: the spread of media content through the blogosphere. In: Proceedings of ICWSM'11, Barcelona, Spain

Chang CC, Lin CJ (2011) Libsvm: a library for support vector machines. ACM Trans Intell Syst Technol 2:27.1–27.27

Chang R, Pimentel S, Svistunov A (2011) Sentiment analysis of occupy wall street tweets. http://cs229.stanford.edu/proj2011/ChangPimentelSvistunov-SentimentAnalysisOfOccupyWallStreet Tweets.pdf

Chen H, Zimbra D (2010) Ai and opinion mining. IEEE Intell Syst 25:74–80

Chmiel A, Sienkiewicz J, Thelwall M, Paltoglou G, Buckley K, Kappas A (2011a) Collective emotions online and their influence on community life. PLoS ONE 6(7):e22207

Chmiel A, Sobkowicz P, Sienkiewicz J, Paltoglou G, Buckley K, Thelwall M (2011b) Negative emotions boost user activity at BBC forum. Phys A 390(16):2936–2944

Dalgleish T, Power M (1999) Handbook of cognition and emotion. Wiley, New York

Diakopoulos NA, Shamma DA (2010) Characterizing debate performance via aggregated Twitter sentiment. In: Proceedings of CHI'10, Atlanta, GA, pp 1195–1198

Ding X, Liu B, Yu PS (2008) A holistic lexicon-based approach to opinion mining. In: Proceedings of WSDM'08, Palo Alto, CA, pp 231–240

Dodds P, Danforth C (2009) Measuring the happiness of large-scale written expression: songs, blogs, and presidents. J Happiness Stud. doi:10.1007/s10902-009-9150-9

Fan RE, Chang KW, Hsieh CJ, Wang XR, Lin CJ (2008) LIBLINEAR: a library for large linear classification. J Mach Learn Res 9:1871–1874

Fouche G (2011) Nobel peace prize may recognise Arab spring. (28 Sept, 2011, Reuters). http://in.reuters.com/article/2011/09/27/idININdia-59582020110927. Accessed 27 Dec 2011

Gazzar SE, Vitorovich L, Bender R (2011) Egypt communications cut ahead of further protests (28 Jan 2011, The Wall Street Journal). http://online.wsj.com/article/BT-CO-20110128-706943.html. Accessed 27 Dec 2011

Godbole N, Srinivasaiah M, Skiena S (2007) Large-scale sentiment analysis for news and blogs. In: Proceedings of ICWSM'07, Boulder, CO

Gonzalez-Bailon S, Paltoglou G (2012) The positive effects of negative emotions in online communities (under review)

Gonzalez-Bailon S, Banchs RE, Kaltenbrunner A (2010) Emotional reactions and the pulse of public opinion: Measuring the impact of political events on the sentiment of online discussions. CoRR. http://arxiv.org/abs/1009.4019

Hall M, Frank E, Holmes G, Pfahringer B, Reutemann P, Witten IH (2009) The weka data mining software: an update. SIGKDD Explor Newsl 11:10–18

Halliday J, Garside J (2011) Rioting leads to cameron call for social media clampdown (11 Aug, 2011, The Guardian). http://www.guardian.co.uk/uk/2011/aug/11/cameron-call-social-media-clampdown. Accessed 27 Dec 2011

Harvey M (2010) Facebook ousts Google in us popularity (17 Mar 2010, The Sunday Times). http://technology.timeson-line.co.uk/tol/news/tech_and_web/the_web/article7064973.ece. Accessed 05 July 2010

Howard P (2011) The Arab spring's cascading effects (23 Feb 2011, Miller-McCune). http://www.miller-mccune.com/politics/the-cascading-effects-of-the-arab-spring-28575/. Accessed 27 Dec 2011

Jijkoun V, de Rijke M, Weerkamp W (2010) Generating focused topic-specific sentiment lexicons. In: Proceedings of ACL '08, Columbus, OH, pp 585–594

Joachims T (1999) Making large-scale SVM learning practical. In: Advances in kernel methods - support vector learning, vol 11. MIT Press, Cambridge, MA

John GH, Langley P (1995) Estimating continuous distributions in Bayesian classifiers. Stanford University, Stanford, CA, pp 338–345

Kramer AD (2010) An unobtrusive behavioral model of "gross national happiness". In: Proceedings of CHI'10, Atlanta, GA, pp 287–290

Le Cessie S, Van Houwelingen JC (1992) Ridge estimators in logistic regression. Appl Stat 41 (1):191–201

Manning CD, Schütze H (1999) Foundations of statistical natural language processing. MIT Press, Cambridge, MA

Mauss IB, Robinson MD (2009) Measures of emotion: a review. Cogn Emotion 23(2):209–237

McCallum AK (2002) Mallet: a machine learning for language toolkit. http://www.cs.umass.edu/mccallum/mallet

Miller GA (1995) Wordnet: a lexical database for English. Commun ACM 38(11):39–41

Mishne G (2005) Experiments with mood classification in blog posts. In: 1st Workshop on stylistic analysis of text for information access, Salvador, Brazil

Mishne G, de Rijke M (2006) Capturing global mood levels using blog posts. In: Proceedings of AAAI-CAAW, Stanford University, Stanford, CA, pp 145–152

Mitchell TM (1997) Machine learning, 1st edn. McGraw-Hill, New York

Mitrović M, Paltoglou G, Tadić B (2011) Quantitative analysis of bloggers' collective behavior powered by emotions. J Stat Mech Theory Exp 2011(02):P02005

O'Connor B, Balasubramanyan R, Routledge BR, Smith NA (2010) From tweets to polls: linking text sentiment to public opinion time series. In: Proceedings of ICWSM'10, Washington, DC

Owsley S, Sood S, Hammond KJ (2006) Domain specific affective classification of documents. In: Proceedings of AAAICAAW'06, Stanford University, Stanford, CA, pp 181–183

Pak A, Paroubek P (2010) Twitter as a corpus for sentiment analysis and opinion mining. In: Proceedings of LREC'10, Valletta, Malta

Paltoglou G, Thelwall M (2012) Twitter, myspace, digg: unsupervised sentiment analysis in social media. ACM TIST 3(4):66.1–66.19

Paltoglou G, Thelwall M, Buckely K (2010) Online textual communication annotated with grades of emotion strength. In: Proceedings of EMOTION, Imperial College, London, pp 25–31

Paltoglou G, Theunis M, Kappas A, Thelwall M (2013) Predicting emotional responses to long informal text. J IEEE Trans Affect Comput 99(PrePrints):1

Pang B, Lee L (2008) Opinion mining and sentiment analysis. Found Trends Inform Retrieval 2(1–2):1–135

Pang B, Lee L, Vaithyanathan S (2002) Thumbs up? Sentiment classification using machine learning techniques. In: Proceedings of EMNLP'02. Association for Computational Linguistics, Philadelphia, pp 79–86

Pennebaker JW (2008) Debate 3: Mccain and Obama word usage (15 Oct 2008, WordWathcers). http://wordwatchers.wordpress.com/2008/10/15/debate-3-mccain-and-obama-word-usage/. Accessed 01 Feb 2012

Pennebaker JW, Francis ME (1999) Linguistic inquiry and word count, 1st edn. Lawrence Erlbaum, Mahwah, NJ

Pennebaker JW, Persaud R (2010) The 2010 UK election: the second debate (23 Apr 2010, WordWathcers). http://wordwatchers.wordpress.com/2010/04/23/the-2010-uk-election-the-second-debate/. Accessed 01 Feb 2012

Platt, J. C. (1999). Fast training of support vector machines using sequential minimal optimization. John C. Platt Microsoft Research 1 Microsoft Way, Redmond, WA, pp 185–208

Quirk R, Greenbaum S, Leech G, Svartvik J (1985) A comprehensive grammar of the English language. Longman, London

Russell JA (1980) A circumplex model of affect. J Person Soc Psychol 39(6):1161–1178

Sebastiani F (2002) Machine learning in automated text categorization. ACM Comput Surv 34(1):1–47

Strapparava C, Mihalcea R (2007) Semeval-2007 task 14: affective text. In: Proceedings of SemEval'07, Prague, Czech Republic, pp 70–74

Strapparava C, Mihalcea R (2008) Learning to identify emotions in text. In: Proceedings of SAC'08, Fortaleza, Ceara, Brazil, pp 1556–1560

Strapparava C, Valitutti A (2004) WordNet-Affect: an affective extension of WordNet. In: Proceedings of LREC'04 (Vol 4), Lisbon, Portugal, pp 1083–1086

Taboada M, Brooke J, Tofiloski M, Voll K, Stede M (2001) Lexicon-based methods for sentiment analysis. Comput Linguistics 37:267–307

Thelwall M (2009) Myspace comments. Online Inform Rev 33(1):58–76

Thelwall M, Buckley K, Paltoglou G, Di C, Kappas A (2010) Sentiment strength detection in short informal text. JASIST 61(12):2544–2558

Thelwall M, Buckley K, Paltoglou G (2011) Sentiment in Twitter events. J Am Soc Inf Sci Technol 62:406–418

Thomas M, Pang B, Lee L (2006) Get out the vote: determining support or opposition from congressional floor-debate transcripts. In: Proceedings of EMNLP'06. Association for Computational Linguistics, Morristown, pp 327–335

Tonkin E, Pfeiffer HD, Tourte G (2012) Twitter, information sharing and the London riots. Bull Am Soc Inf Sci Tech 38(2):49–57

Vapnik VN (1999) The nature of statistical learning theory (information science and statistics). Springer, Heidelberg

Velikovich L, Blair-Goldensohn S, Hannan K, McDonald R (2010) The viability of web-derived polarity lexicons. In: Proceedings of HLT'10, Stroudsburg, PA, USA, pp 777–785

Whitelaw C, Garg N, Argamon S (2005) Using appraisal groups for sentiment analysis. In: Proceedings of CIKM'05, Bremen, Germany, pp 625–631

Wiebe J, Bruce RF, O'Hara TP (1999) Development and use of a gold-standard data set for subjectivity classifications. In: Proceedings of 37th annual meeting of ACL, College Park, MD, USA, pp 246–253

Wilson T, Hoffmann P, Somasundaran S, Kessler J, Wiebe J, Choi Y (2005a) Opinionfinder: a system for subjectivity analysis. In: Proceedings of HLT/EMNLP-demo, Vancouver, BC, Canada, pp 34–35

Wilson T, Wiebe J, Hoffmann P (2005b) Recognizing contextual polarity in phrase-level sentiment analysis. In: Proceedings of EMNLP'05, Vancouver, BC, Canada, pp 347–354

Zaidan O, Eisner J, Piatko CD (2007) Using "annotator rationales" to improve machine learning for text categorization. In: Proceedings of HLT-NAACL. Association of Computational Linguistics, Rochester, NY, pp 260–267

Emotion Analysis on Social Media: Natural Language Processing Approaches and Applications

Dipankar Das and Sivaji Bandyopadhyay

Abstract The rapidly growing online activities in the Web motivate us to analyze the reactions of different emotional catalysts on various social networking substrates. Thus, in the present chapter, different concepts, motivations, approaches and applications of emotion analysis are discussed in order to achieve the main challenging tasks such as feature representation schema, emotion classification, holder and topic detection and identifying their co-references, etc. as these are the main salient points to cover while analyzing emotions in social media. Additionally, a prototype is also described and assessed to analyze emotions, its collective actions based on users and topics, its components and their association from different available data sets of English and Bengali as case studies. Experiments and final outcomes highlight the promise of the approach and some open research problems.

1 Introduction

In recent times, research activities in the areas of opinion, sentiment and/or emotion in natural language texts and other media are gaining ground under the umbrella of Subjectivity Analysis and Affective Computing. Subjectivity Analysis aims to identify whether a sentence expresses an opinion or not and if so, whether the opinion is positive or negative (Liu 2010). The emotions are the subjective feelings and thoughts, and the strengths of opinions are closely related to the intensities of

D. Das (✉)
Computer Science & Engineering Department, Jadavpur University, Kolkata, India
e-mail: dipankar.dipnil2005@gmail.com

S. Bandyopadhyay
Computer Science & Engineering Department, Jadavpur University, Kolkata, India
e-mail: sivaji_cse_ju@yahoo.com

N. Agarwal et al. (eds.), *Online Collective Action*, Lecture Notes in Social Networks,
DOI 10.1007/978-3-7091-1340-0_2, © Springer-Verlag Wien 2014

certain emotions, e.g., joy and anger (Liu 2010). Though the concepts of emotions and opinions are not equivalent, they have a large intersection. On the other hand, emotion analysis has also been studied in many fields of Affective Computing, a key area of research in computer science. The majority of the emotion analysis fields are hermeneutics, psychology, philosophy, sociology, biology and political science.

The 24/7 news sites and blogs facilitate the expression and shaping of opinion or emotion locally and globally (Ahmad 2011). Emails, Weblogs, chat rooms, online forums and even Twitter are being considered as the effective social media for discussing recent topics. Blogs are the most important, communicative and informative repository of text-based emotional contents in Web 2.0 (Yang et al. 2007). Facebook, Linkdin and even Google + also contain a blog-like structure. Many blogs act as online diaries of the bloggers for reporting the blogger's daily activities and surroundings. Sometimes, the blog posts are annotated by other bloggers. Therefore, blogs are being considered as one of the personal journals where people express their personal opinions on different aspects like products, travelled tourism places, politics and current happenings in society. The blog posts contain instant views, updated views or influenced views regarding single or multiple topics.

Nowadays, people post the product reviews at merchant sites and express their views on discussion forums, blogs and social network sites. But, with huge amounts of such social text being generated, it is important to find methods that can annotate and organize documents in meaningful ways. Thus, topic identification is also used for document ranking in information retrieval systems. In addition to the content of the document itself, other relevant information about a document such as related topics can often enable a faster and more effective search or classification.

Topic identification from social sites is also essential in connection with categorizing search applications (Stein and Eissen 2004). Categorizing search means to apply text categorization facilities to retrieval tasks where a large number of documents are returned. Categorizing search has attracted much interest recently; its potential has been realized by users and search engine developers in the same way. The categorization was also started based on the opinionated or sentiment or emotional contents of the documents in the World Wide Web since the last few years.[1,2]

From the perspective of natural language processing (NLP) applications, emotion analysis has been considered as a sub-discipline at the crossroads of information retrieval (Pang and Lee 2008) (Sood and Vasserman 2009) and computational linguistics (Wiebe et al. 2005). Emotions, of course, are not linguistic things. However, the most convenient access that we have to them is through language (Strapparava and Valitutti 2004). Natural language texts not only contain informative contents, but also more or less attitudinal private information including emotions. It has been observed that the classification of reviews (Turney 2002) or

---

[1] http://twittersentiment.appspot.com/

[2] http://www.tweetfeel.com/

newspaper articles (Lin et al. 2007) is also increasingly incorporating emotion analysis within their scope. But the identification of emotions from texts is not an easy task because of its restricted access in case of objective observation or verification (Quirk et al. 1985). Moreover, the same textual content can be presented with different emotional slants (Grefenstette et al. 2004).

It is said that sentiment or emotion is typically a localized phenomenon that is more appropriately computed at the paragraph, sentence or entity level (Liu 2009). But emotion analysis can be performed at several levels of granularity, word, phrase, sentence, clause, paragraph or document (Das and Bandyopadhyay 2009, 2010a). Yu (2009) also proposed several granularity levels such as terms, expression, statement, passage and document.

It is sometimes observed that the topics discussed at the sentence level are not similar to the topic of the overall document. Thus, it is important to find methods that can annotate the documents in meaningful ways so that the topic identification can also be used for document ranking in information retrieval systems. One of the important tasks that proposed various insights and solutions related to the topic identification was described in Lin (1997). Paula Chesley et al. (2006) present experiments on subjectivity and polarity classifications of topic- and genre-independent blog posts. Emotion analysis also involves identifying the emotion holder in addition to the emotion topic. An emotion agent or holder is the person or organization expressing the emotion (Wiebe et al. 2005). In the case of product reviews and blogs, holders are usually the authors of the posts. Extraction of the emotion holder is important to discriminate between emotions that are viewed from different perspectives (Seki 2007). By grouping emotion holders of different stances on diverse social and political issues, we can have a better understanding of the relationships among countries or among organizations (Kim and Hovy 2006).

Sometimes, it is found that the sentences of a document may or may not contain any direct clue for the emotional expression, especially in social sites ("Dream of music is in their eyes and hearts".) or there are certain example sentences that contain emotional expression without a potential holder ("His acting was really attractive"), or topic ("I fall into cry"). Thus, with such examples and problems in mind, it can be hypothesized that the notion of user-topic co-reference will facilitate both the manual and automatic identification of emotions. Therefore, we consider the emotional expression, holder and topic as the three essential components of emotion (Das and Bandyopadhyay 2010d).

The tracking of emotions over events or about politics as expressed in online forums or news to customer relationship management and the determination of the emotion holder and topic is an important task. Thus, we can also track users' emotions expressed in online forums or blogs or Twitter messages or social networking sites for different applications such as sentiment review, customer management, stock exchange prediction, etc. The identification of the temporal trends of emotions and topics has drawn the recent attention of NLP communities (Fukuhara et al. 2007; Das and Bandyopadhyay 2011a) because among all concerns, emotions of people are important because people's emotion has great influence on our society (Das and Bandyopadhyay 2011b).

Apart from the commercial perspectives, the other potential contributions of the present chapter in relation to the book have been described as follows:

- We aim to identify the emotional changes among users during their communication in the context of social networking. We mainly focus on the blog users and their topics of discussion for identifying emotional changes with respect to time. We have considered the social interactions of the bloggers as collective actions, and their emotional changes based on such actions have been measured from the perspectives of topic and time.
- We have incorporated the knowledge of two types of stimuli, self and influential affects in identifying the behavioural contingencies from the social interactions of the blog users.
- We consider collective actions as events, and therefore the emotional change with respect to events has been represented using the graphical notion of sentiment event tracking.

Thus, based on the above issues, we have organized this chapter as follows: Section 2 describes the concepts and motivations related to emotion analysis. Section 3 discusses a prototype for emotion analysis which identifies several components of emotion and the need of emotional co-reference for the components. Section 4 describes the application of the system in terms of time-based emotion tracking. Finally, Section 5 highlights the main conclusions of this chapter.

2   Concepts and Motivations

Several frameworks exist from various fields of academic study, such as cognitive science, linguistics and psychology, that can inform and augment analyses of sentiment, opinion and emotion (Read and Carroll 2010). Some of the generalized definitions of emotions are as follows:

Definition[1]: Emotion is a complex psycho-physiological experience of an individual's state of mind as interacting with biochemical (internal) and environmental (external) influences. In humans, emotion fundamentally involves physiological arousal, expressive behaviours and conscious experience (Myers 2004).

Definition[2]: In psychology and common use, emotion is an aspect of a person's mental state of being, normally based on or tied to the person's internal (physical) and external (social) sensory feeling (Zhang et al. 2008).

Emotions, of course, are not linguistic objects/entities, and it is also said that the most non-phenomenological access to emotion is language (Ortony and Turner 1990). But the identification of emotions from texts is not an easy task due to the following challenges:

- Basic and complex categories of emotions (James 1884; McDougall 1926; Watson 1930; Arnold 1960; Izard 1971; Plutchik 1980; Ekman 1992; Parrott 2001 and several others)

Though there are several other theories of emotion, the debate is concerned with some basic and complex categories, where complex emotions could arise from cultural conditioning or association combined with basic emotions. However, there is still not a set of agreed basic emotions of people among researchers. Ekman (1992), for instance, derived a list of six basic emotions from subjects' facial expressions which several researchers employed as classes in an affect recognition task. Thus, we have presently confined ourselves into these six classes of emotions.

- Restricted access in case of objective observation or verification (Quirk et al. 1985)
- Presentation of the same textual content with different emotional slants (Grefenstette et al. 2004)

Although there are only 6 forms of emotions, there are a large number of language expressions that can be used to express them.

- Level of granularity for processing of emotional evaluative expressions (word/ phrase/clause/sentence/paragraph/document) (Wiebe et al. 2005; Das and Bandyopadhyay 2009, 2010a; Liu 2009; Yu 2009)
- Valence (*positive, negative, neutral*) or Ekman's six emotion types— *"anger"*, *"disgust"*, *"fear"*, *"joy"*, *"sadness"*, *"surprise"*
- *Intensity* (low, medium, high, etc.)
- Aspects and attributes (Holder/Target/Topic) (Kim and Hovy 2006; Das and Bandyopadhyay 2010b, c)

Thus, it is very hard to define emotion and to identify its regulating or controlling factors. These aspects raise the need of syntactic, semantic and pragmatic analysis of a text (Polanyi and Zaenen 2006).

But the majority of subjective analysis methods that are related to emotion are based on textual keywords spotting that uses specific lexical resources. SentiWordNet (Baccianella et al. 2010) is a lexical resource that assigns positive, negative and objective scores to each WordNet synset (Miller 1995). Subjectivity wordlist (Banea et al. 2008) assigns words with strong or weak subjectivity and prior polarities of types positive, negative and neutral. Some well-known sentiment lexicons have been developed, such as General Inquirer System (Stone 1966), Subjective adjective list (Baroni and Vegnaduzzo 2004), English SentiWordNet (Esuli and Sebastiani 2006), Taboada's adjective list (Voll and Taboada 2007), etc. But all the mentioned resources are in English and have been used in coarse-grained sentiment analysis (e.g., positive, negative or neutral). The characterization of the words and phrases according to their emotive tone was first carried out in Turney (2002).

In recent trends, the application of mechanical turk for generating emotion lexicons (Mohammad and Turney 2010) shows promising results. The opinion or emotion annotation of a language has been performed for several natural language domains such as news (Strapparava and Mihalcea 2007), blogs (Mishne and de Rijke 2006a, b) or others. Opinion mining at word, sentence and document levels along with opinion summarization on news and Weblog documents is discussed in Ku et al. (2006). In order to estimate the affects in text, the model proposed in Neviarouskaya et al. (2007) processes symbolic cues and employs NLP techniques for word, phrase and sentence level analysis. Several machine-learning techniques were employed on blog data to identify the mood of the authors during writing (Mishne and de Rijke 2006a, b) The text-based emotion prediction using such supervised machine-learning approaches based on the SNoW learning architecture is discussed in Alm et al. (2005).

Prior work in identification of opinion or emotion holders has sometimes identified only a single opinion per sentence (Bethard et al. 2004) and sometimes several (Choi et al. 2005). An identification of opinion holders for question answering with supporting annotation tasks was attempted in Wiebe et al. (2005). The techniques that were employed to detect the holders are based on labelling the arguments of the verbs with their semantic roles (Swier and Stevenson 2004) or syntactic models (Das and Bandyopadhyay 2010b), emotion knowledge base (Hu et al. 2006), machine–learning-based classification (Evans 2007), etc. The anaphor resolution–based opinion holder identification method also exploits the lexical and syntactic information (Kim et al. 2007).

On the other hand, emotion topic can be defined as the real-world object, event or abstract entity that is the primary subject of the emotion or opinion as intended by its holder (Stoyanov and Cardie 2008a). The topic depends on the context in which its associated emotional expression occurs (Stoyanov and Cardie 2008b). In the related area of opinion topic extraction, different researchers have contributed their efforts. Some of the works are mentioned in Kobayashi et al. (2004), Popescu and Etzioni (2005). But all these works are either based on lexicon lookup or are applied on the domain of product reviews. The opinion topics are not necessarily spatially coherent in social texts as there may be two opinions in the same sentence on different topics as well as opinions that are on the same topic separated by opinions that do not share that topic (Stoyanov and Cardie 2008a). Not only topics, but the relation between sentiment and event can also be identified using the knowledge of lexical equivalence and co-reference approaches (Kolya et al. 2011).

The majority of the existing works in this field have been conducted for English. As far as our discussions on related work mentioned earlier are concerned, to the best of our knowledge, at present, only a few corpora or systems are available for analyzing emotions in languages other than English. In the present chapter, we have considered the Bengali language for our case studies. Bengali is the fifth most popular language in the world, second in India and the national language in Bangladesh. Hence, the present work is also a foray into emotion analysis for an Indian language.

It can be concluded that the perspectives of sociology, psychology and commerce along with the close association among people, topics and sentiments motivate us to investigate the insides in emotional changes of people over topic and time.

3 Approaches to Identifying Emotion Components

There are two main approaches to tackle the issues of emotion analysis: A "rule-based" approach defines markers and linguistic syntactic rules so as to determine the emotions of a text (Mihalcea et al. 2007). It also addresses the semantics associated with emotions by using conceptual resources such as lexicons. On the other hand, a "corpus-based" approach uses an annotated corpus to build a system which identifies emotions. In the corpus-driven approach, the language choice only impacts the corpus selection. However, most successful approaches depend on syntactic rules or text semantics of either keywords or phrases.

3.1 Emotional Expression

It is said that sentiment and/or emotion is typically a localized phenomenon that is more appropriately computed at the paragraph, sentence or entity level (Liu 2009). But emotion analysis can be performed at several levels of granularity, from word, phrase and sentence to document levels. Thus, we can propose the prototype systems of identifying evaluative emotional expressions at different levels of granularities such as word (W), phrase (P), sentence (S) and document (D) (Das and Bandyopadhyay 2010a, 2010f; Das and Bandyopadhyay 2011c).

The baseline system for word-level emotion tagging to measure the performance with respect to each emotion class has been developed in Das and Bandyopadhyay (2010a). Each of the words of the corpus is passed through these six separate modules to tag with the appropriate class label. In addition, the conditional random field (CRF) (Lafferty et al. 2001)– and support vector machine (SVM) (Joachims 1998)–based machine-learning classifiers are also employed for word-level emotion tagging. Different singleton features (e.g., part of speech (POS) of the words, question words, reduplication, colloquial/foreign words, special punctuation symbols, negations, emoticons) and context features (e.g., unigram, bigram) at the word and POS tag level along with their different combinations are used for training and testing. Some of the common features are as follows:

- **Bag of Words:** Considers all terms in the corpus and builds a vector per document, where each dimension expresses the presence or absence of the term in the document. This approach can be refined adding the term importance in the corpus.
- **Emotion/Affect Words (EW):** The presence of a word in the WordNet Affect lists (Strapparava and Valitutti 2004) identifies the emotion/affect words.

- **Parts of Speech (POS):** We are interested in the verb, noun, adjective and adverb words as these are emotion-informative constituents.
- **Bag of features:** Some features are chosen, e.g., some units with high frequency in the corpus, such as unigrams, n-grams, POS, adjectives, etc. In order to represent it, usually each feature is put in a dimension of a vector representing the text fragment. In each dimension, usually only the feature presence or frequency is recorded.
- **Intensifiers (INTF):** Dependency relations such as amod() [adjectival modifier] and advmod() [adverbial modifier] containing JJ (adjective) and RB (adverb) tagged elements are considered as intensifiers.
- **Direct and Transitive Dependency relations (DD and TD):** The direct dependency (DD) is identified based on the simultaneous presence of the emotion word and the other word in the same dependency relation, whereas the transitive dependencies (TD) are verified if they are connected via one or more intermediate dependency relations.
- **Negations (NEG):** Dependency relations such as neg_( ) [negation modifier] or the negative words from a manually prepared list (no, not, nor, neither, etc.), e.g., "there's no way I'm turning it down", are adopted for the Negations feature.
- **Conjuncts (CONJ):** The dependency relation conj_( ) [conjunct] identifies the Conjuncts features (and, or, but), e.g., "But I asked him about WHY he doesn't cook for himself".
- **Punctuation Symbols (Sym):** Symbols such as comma (,), (!), (?) are often used in single or multiple numbers to emphasize emotional expressions and considered as crucial clues for identifying emotional presence ("I can't believe she is FINALLY here!!!").
- **Discourse Markers (DM):** The present task aims to identify only the explicit discourse markers that are tagged by conjunctive_() or mark_() type dependency relations of the parsed constituents (e.g., as, because, while, whereas).
- **Capitalized Phrases (CP):** A capitalized word or a long capitalized phrase segment, e.g., "I forgot how demeaning BME classes are" or "Terrorists MAKE ME SICK, they ought to all be horrifically detained", is considered as the Capitalized Phrases feature.
- **Emoticons (emot_icon):** The emoticons (☺, ☹, ☺) and their consecutive occurrence generally contribute as much as real sentiment to the words that precede or follow them.

Lexical analysis plays a crucial role to identify emotions from a text. For example, the words like *love*, *hate*, *good*, *bad*, *happy* and *sad* directly indicate emotion. But the assumption is true only within a limited context and restricted granularity. Let us consider the following example:

Example 1 *"Though Mr. Jonathon Read could be generally **happy** about his car, his wife might be **dissatisfied** by the engine noise".*

If the sentence would be written by exchanging only the positions of the direct sentiment words (**happy** and **dissatisfied**), the whole sentence might lead to a mess:

"Though Mr. Jonathon Read could be generally **dissatisfied** about his car, his wife might be **happy** by the engine noise".

Thus, the prime factors for disambiguation coming to our mind are the components that are associated with the emotional expressions. In the example, **Mr. Jonathon Read** and **his car** are associated with the emotional expression **happy,** whereas **his wife** and **engine noise** are associated with the emotional expression **dissatisfied.** It is clear that **Mr. Jonathon Read** and **his wife** represent the emotion holders, whereas **his car** and **engine noise** represent the emotion topics, respectively.

3.2 Emotion Holder

The baseline model (BM) for identifying emotion holders in English can be designed based on the subject information of the parsed emotional sentences. We can employ the Stanford dependency parser[3] to accomplish the task. Similarly, for the morphologically rich languages, e.g., Bengali, the sentences are passed through an open source Bengali shallow parser that produces different morphological information (e.g., root, case, vibhakti, tam, suffixes, etc.). The lexical pattern–based phrase-level similarity clues containing different POS combinations, named entities (NEs) and noun phrases are considered for identifying the emotion holders.

Emotion holders can also be identified based on the syntactic argument structure of the emotional sentences. In English, the head of each chunk in the dependency-parsed output helps in constructing the syntactic argument structure with respect to the key emotional verb. Two separate techniques can be used for extracting the argument structure. One is from the parsed result directly, and another is from the corpus that has been POS-tagged and chunked separately. Similarly, the verb-based argument structures are acquired from the chunk- or phrase-level lexical patterns (Kim and Hovy 2006; Choi et al. 2005; Das and Bandyopadhyay 2010b). The pivotal hypothesis considered in the syntactic model (SynM) is based on the hypothesis followed in Banerjee et al. (2010). The hypothesis is that if the acquired syntactic argument structure of a sentence matches with any of the retrieved frame syntaxes of VerbNet,[4] the holder roles (e.g., *Experiencer, Agent, Actor, Beneficiary,* etc.) associated with the VerbNet frame syntaxes are then assigned in the appropriate slots in the syntactic arguments of the sentence. For other languages, each acquired syntactic argument structure is mapped to all the possible frame syntaxes present for the corresponding verb in the English VerbNet.

---

[3] http://nlp.stanford.edu/software/lex-parser.shtml

[4] http://verbs.colorado.edu/~mpalmer/projects/verbnet.html

3.3   Emotion Topic

Like emotion holders, the baseline model for identifying emotion topics is developed based on the object-related dependency-parsed relations. The phrase segments containing topic-related *thematic roles* (e.g., *Topic*, *Theme*, *Event*, etc.) are extracted from the verb-based syntactic argument structures of the sentences. On the other hand, a supervised model (SvdM) is adopted to identify multiple emotion topics along with their individual topic and target spans from each sentence. CRF, SVM and Fuzzy Classifier (FC) are employed by considering various features (e.g., the annotated emotional expressions along with direct and transitive dependencies, causal verbs, discourse markers, emotion holders, named entities and four types of similarity measures—Structural Similarity, Sentiment Similarity, Syntactic Similarity and Semantic Similarity) and their combinations (Kim and Hovy 2006; Das and Bandyopadhyay 2010c). A prototype system for identifying sentiment has been shown in Fig. 1 taken from the article (Kim and Hovy 2006). The incorporation of a special feature, Structural Similarity, that is based on the Rhetorical Structure Theory (Mann and Thompson 1988) improves the topic identification system.

3.4   Need of Emotion Co-reference

The importance of the emotion-associated components such as holder and topic can easily be verified by mingling their positions and keeping the positions of their corresponding emotional expressions intact. In Example 1, the following variations can be seen:

1. *"Though **his wife** could be generally **happy** about **his car**, **Mr. Jonathon Read** might be **dissatisfied** by **the engine noise**"*.
2. *"Though **Mr. Jonathon Read** could be generally **happy** about **the engine noise**, **his wife** might be **dissatisfied** by **his car**"*.
3. *"Though **his wife** could be generally **happy** about **the engine noise**, **Mr. Jonathon Read** might be **dissatisfied** by **his car**"*.

Thus, it is clear that the proper understanding of the emotion components and their associations is very important if we need to mine emotion properly from texts. For this very reason, document-level emotion classification sometimes fails to detect emotion as it does not account for the individual emotion components. Sometimes, a single topic is co-referred by several users as well as multiple topics are co-referred by a single user. Ekman's 6 different emotions plotted for 8 different topics and referred by each of the 22 bloggers are shown in Fig. 2 signifying that the user-topic co-reference system performs to generate the emotional views of the bloggers and its dependence on the associated topics. With the above examples and problems in mind, it can be hypothesized that the co-reference among the emotion

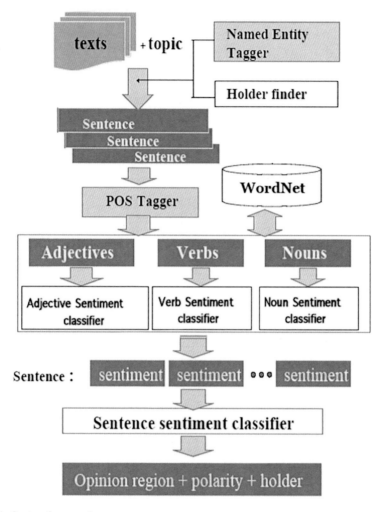

Fig. 1 System framework

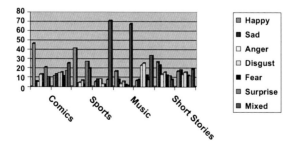

Fig. 2 Topic-based emotions (in %) of the blog users

components will facilitate both the manual and automatic identification of emotions.

3.5 Results

Various experiments regarding symbolic feature, language, and domain-dependent features are carried out for evaluating the word-level emotion classification system. The lexical feature (e.g., POS, words of SentiWordNet and WordNet Affect) outperforms other features significantly. A different combination of context features also shows significant improvement in performance. Though we evaluated all our systems on large amounts of data and the corresponding results are mentioned in different research articles (Das and Bandyopadhyay 2009, 2010a, b, c, d, e, f, 2011b; Das and Bandyopadhyay 2011c), we have reported here only the best performed results on small test data as follows:

The word-level tagging system has demonstrated the average F-Scores of 83.65 % on 1,500 word tokens on English news corpus. The average F-Score of 65 % was achieved for 200 test sentences of SemEval 2007 corpus in sentential emotion tagging. A supervised system for English blogs (Aman and Szpakowicz 2007) outperforms the baseline system and achieves the average F-Scores of 82.72 %, 76.74 % and 89.21 % for emotional expressions, sentential emotions and intensities respectively on 565 gold standard test sentences.

It has been observed that the baseline model for English suffers in identifying emotion holders from the passive sentences. The dependency parser–based method achieves a better F-Score (66.98 %) than the other method (F-Score of 62.39 %) on a collection of 4,112 emotional sentences as the second method fails to disambiguate mostly the arguments from adjuncts. The maximum average F-Scores of the baseline and hybrid systems for emotion topic identification are 56.75 % and 58.88 % respectively on 500 sentences (Table 1). The supervised multi-engine voting system achieves the F-Scores of 70.51 % and 90.44 % for topic and target span identification respectively from the blog sentences. But the syntactic system suffers in resolving some errors (e.g., appositive cases, co-reference with emotional expression, multiple holders and topics, overlapping topic spans, anaphoric presence of the holders). Thus, some simple rule-based error reduction techniques based on rhetorical structure and emotional expressions are employed in the syntactic system.

It was observed that the error occurs mostly for metaphoric and colloquial usages, unstructured sentences (e.g., "Really starting to lose it") and sentences containing typographic errors (e.g., "she's feeling very goooood about herself"). But it is also true that in micro-blogging such as Twitter, the number of misspelled words is even higher due to the 140-character space constraint. In order to test the robustness of the proposed approaches, we would like to incorporate the approaches in order to annotate a Twitter-extracted corpus.

Table 1 F-Scores (in %) of three emotional components

Topics	English	Bengali
Evaluative Expressions	83.65 (W, #1500) ; 76.74 (S, #565); 82.72 (P, #603)	70.23 (W, #1500); 82.72 (S, #200);76.74(D, #110)
Emotion Holder	64.83 (BM) (# 4112) 66.98 (SynM) (# 4112)	53.85 (BM) (#500) 66.03 (SynM) (#500)
Emotion Topic	56.75 (BM) (#500) 70.51(SvdM) (#500)	50.02 (BM) (#500) 61.98 (SynM) (#500)

4 Application in Emotion Tracking

4.1 Tracking Bloggers' Emotions

The blog documents are generally stored in the format as shown in Fig. 3. Each of the blog documents is assigned with a unique identifier (docid#) followed by a section devoted for topics and several sections devoted for different users' comments. Each comment section consists of several nested and overlapped subsections that also contain the bloggers' comments. Each of the comment sections of an individual blogger is uniquely identified by the notion of section identification number (secid#). Each section contains the information regarding the identification number of the blog user (uid#) and the associated timestamp (tid#).

If we consider the individual comment section as a separate paragraph that contains several emotional sentences, the emotions present in such individual comment sections represent the emotional state of the blogger at that timestamp. The Referential Informative Chain (RIC) for each of the bloggers is constructed by acquiring the default annotated information like timestamp (tid#), unique identifier (uid#) and emotional comments that are acquired from the nested tree-like structure of the comment sections. The individual RIC is developed for each single blogger with respect to each comment section. Each node of an RIC denotes the emotional state of the blogger at a particular time instance, and the sequence of adding information into the nodes is based on the ascending order of associated timestamps. For example, in Fig. 2, the two nodes, namely, $n1$ and $n2$, will be added in front of an RIC developed for the blog user with $uid = 1$. The associated timestamps ($t1$, $t4$) and emotions will also be added into the nodes accordingly. As $t4 > t1$, the inclusion of node $n1$ is considered before the inclusion of $n2$ into the corresponding RIC of uid 1.

The identification of Ekman's six basic emotions from the bloggers' comments is carried out at sentence- and paragraph-level granularities. An affect-scoring technique is employed to identify the emotions of a state or node in each of the RICs of the bloggers. Two types of affect scores, Self Affect Score (SAS) and Influential Affect Score (IAS), are used to produce the Emotional Score (ES) of a blogger at each node with a particular timestamp (Das and Bandyopadhyay 2011c).

The changes of a blogger's emotions are tracked based on the emotions that are assigned to the nodes of the blogger's RIC. The importance of self and influential

Fig. 3 General structure of a blog document

```
-<DOC docid = xyz>
    +<Topic>.... </Topic>
    -<User Comments id=UC1>
        -<U uid=1, tid=t1, secid=UC1>....
            -<U uid=2, tid=t2, secid=UC1.1>...</U>
            -<U uid=3, tid=t3, secid=UC1.2>...</U>
                -<U uid=1, tid=t4, secid=UC1.2.1>...</U>

                ....
        </U>
    </User Comments>
    +<User Comments id=UC2>
    +<User Comments id=UC3>
        ...
</DOC>
```

affects in tracking results is evaluated using two evaluation techniques, extrinsic and intrinsic. The system achieves precision (P), recall (R) and F-score of 61.05 %, 69.81 % and 65.13 % respectively in case of extrinsic evaluation and the satisfactory average scores of 0.67 and 0.71 for the nominal alpha (Nα) and interval alpha (Iα) in case of intrinsic evaluation (Das and Bandyopadhyay 2011c).

4.2   Sentiment Event Tracking

The temporal sentiment identification from social events has been carried out in Fukuhara et al. (2007). In their task, the authors analyzed the temporal trends of sentiments and topics from a text archive that has timestamps in Weblog and news articles. The system produces two kinds of graphs, topic graph that shows temporal change of topics associated with a sentiment and sentiment graph that shows temporal change of sentiments associated with a topic. Mishne and de Rijke (2006a, b) also proposed a system, MoodViews, to analyze the temporal change by using 132 sentiments used in LiveJournal. With respect to information visualization, Havre et al. proposed a system called ThemeRiver (Havre et al. 2002) that visualizes thematic flows along with a timeline. The temporal relations between events associated with similar or different types of sentiments and visualization of sentiment flows on events based on temporal expressions were carried out in Das and Bandyopadhyay (2011a) by incorporating the knowledge of temporal relations (e.g., AFTER, BEFORE, OVERLAP). The authors proposed a task to identify the temporal relations between the events that occur in two consecutive sentences and to classify the event pairs into their respective temporal classes. Incorporating sentiment property into the set of other standard features, the proposed system outperforms all the participated state-of-the-art systems of TempEval 2007. The positive or negative coarse-grained sentiments as well as Ekman's six basic universal emotions or fine-grained sentiments are assigned to the events. Based on the temporal relations, the events from each of the documents are represented using a directed graph that shows the shallow path for identifying the sentiment changes over events. An individual separate graph is generated for each of the documents of

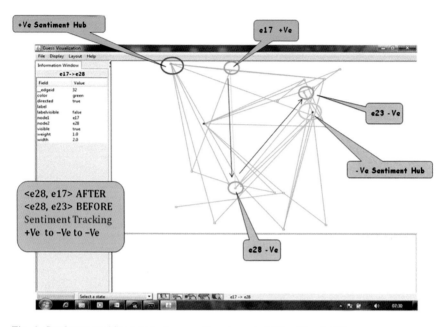

Fig. 4 Sentiment tracking among three sentiment events (*e28*, *e17* and *e23*)

TempEval-2007 event corpus. The sentiment of each sentence is assigned to its containing event, and each event is represented using a graphical node. The event nodes that are of similar sentiments are connected to their corresponding sentiment hubs based on their annotated sentential sentiment tags.

The tracking of sentiments includes the sentiment twist and sentiment transition. The sentiment twist is the change of sentiment between two consecutive events, whereas sentiment transition is among more than two sentiment events (as shown in Fig. 4). The sentiment change or tracking of sentiments is identified from the AFTER, BEFORE and OVERLAP temporal relations. The ambiguities of the OVERLAP relation are identified by the notion of BEFORE-OR-OVERLAP and OVERLAP-OR-AFTER relations. The number of instances of the sentiment transitions is less than the number of instances of the sentiment twists in the TimeML corpus. Hence, the sentiment transition or tracking of sentiment is identified based on the sentiment twists of the intermediate event pairs in an event chain.

5 Conclusions

This chapter gave an introduction to emotion analysis, its foundations and representative state-of-the-art approaches. In addition, the main problems and issues are discussed. Due to many challenging research problems and a wide variety of practical applications, it has been a very active research area in recent years.

We mainly focused on feature-based emotion analysis, which exploits the full power of an abstract model being built. Results of an implemented medium-size prototype are reported in terms of usual metrics such as F-score, accuracy, etc.

Finally, it is important to highlight that all the emotion analysis tasks are very challenging from the perspective of social computing, and our understanding and expertise of the different arisen issues are still limited. The main reason is that it is a natural language processing task, and natural language processing has no easy problems. In addition, the most effective machine-learning algorithms (i.e., CRF, SVM) still produce no human-understandable results, such that, although they may achieve improved accuracy, we know little about how and why, apart from some basic knowledge gained in the manual feature selection process.

Acknowledgments The work reported in this paper was supported by a grant from the India-Japan Cooperative Programme (DSTJST) 2009 Research project entitled "Sentiment Analysis where AI meets Psychology" funded by Department of Science and Technology (DST), Government of India.

References

Ahmad K (2011) Affective computing and sentiment analysis: emotion, metaphor and terminology. Springer text, speech and language technology series, vol 45. Springer, Heidelberg

Alm CO, Roth D, Sproat R (2005) Emotions from text: machine learning for text-based emotion prediction. In: Proceedings of HLT-EMNLP. Association of Computational Linguistics, Stroudsburg, PA, pp 579–586

Aman S, Szpakowicz S (2007) Identifying expressions of emotion in text. In: Matoušek V, Mautner P (eds) Text, speech and dialogue. Lecture notes in computer science 4629. Springer, Heidelberg, pp 196–205

Arnold MB (1960) Emotion and personality. Columbia University Press, New York

Baccianella S, Esuli A, Sebastiani F (2010) SentiWordNet 3.0: an enhanced lexical resource for sentiment analysis and opinion mining. In: Proceedings of the 7th conference on language resources and evaluation, Valleta, Malta, pp 2200–2204

Banea C, Mihalcea R, Wiebe J (2008) A bootstrapping method for building subjectivity lexicons for languages with scarce resources. In: The sixth international conference on language re-sources and evaluation (LREC 2008), Marrakech, Morocco

Banerjee S, Das D, Bandyopadhyay S (2010) Bengali verb subcategorization frame acquisition - a baseline model. In: Proceedings of the 7th workshop of Asian Language Resources (ALR-7), Joint conference of the 47th annual meeting of the association for computational linguistics and the 4th international joint conference on natural language processing of the Asian Federation of Natural Language Processing (ACL-IJCNLP-2009), Suntec, Singapore, pp 76–83

Baroni M, Vegnaduzzo S (2004) Identifying subjective adjectives through web-based mutual information. In: Proceedings of the German Conference on NLP, Vienna

Bethard S, Yu H, Thornton A, Hatzivassiloglou V, Jurafsky D (2004) Automatic extraction of opinion propositions and their holders. In: AAAI Spring symposium on exploring attitude and affect in text: theories and applications. AAAI, Palo Alto, CA

Chesley P, Bruce V, Li X, Srihari RK (2006) Using verbs and adjectives to automatically classify blog sentiment. In: Proceedings of AAAI Spring symposium on computational approaches to analyzing weblogs. AAAI, Palo Alto, CA, pp 25–28

Choi Y, Cardie C, Riloff E, Patwardhan S (2005) Identifying sources of opinions with conditional random fields and extraction patterns. In: Proceedings of HLT/EMNLP, Vancouver, BC, Canada

Das D, Bandyopadhyay S (2009) Word to sentence level emotion tagging for Bengali blogs. In: ACL-IJCNLP 2009, Singapore, pp 149–152

Das D, Bandyopadhyay S (2010a) Sentence level emotion tagging on blog and news corpora. J Intell Syst 19(2):125–134

Das D, Bandyopadhyay S (2010b) Emotion holder for emotional verbs – the role of subject and syntax. In: Gelbukh A (ed) CICLing- 2010. Lecture notes in computer science 6008. Springer, Heidelberg, pp 385–393

Das D, Bandyopadhyay S (2010c) Identifying emotion topic-an unsupervised hybrid approach with rhetorical structure and heuristic classifier. In: Proceedings of the 6th IEEE NLP-KE 2010, Beijing, 21–23 Aug 2010. ISBN 978-1-4244-6896-6

Das D, Bandyopadhyay S (2010d) Discerning emotions of bloggers based on topics – a supervised coreference approach in Bengali. In: Proceedings of the 22nd conference on computational linguistics and speech processing (ROCLING 2010), Taiwan, pp 350–360

Das D, Bandyopadhyay S (2010e) Developing Bengali WordNet affect for analyzing emotion. In: Proceedings of the 23rd international conference on the computer processing of oriental languages (ICCPOL-2010), Redwood City, CA, pp 35–40

Das D, Bandyopadhyay S (2010f) Identifying emotional expressions, intensities and sentential emotion tags using a supervised framework. In: 24th PACLIC, Tohoku University, Sendai

Das D, Bandyopadhyay S (2011a) Emotions on Bengali blog texts: role of holder and topic. First workshop on social network analysis in applications (SNAA 2011). ASONAM 2011:587–592. doi:10.1109/ASONAM.2011.106

Das D, Bandyopadhyay S (2011b) Tracking emotions of bloggers – a case study for Bengali. POLIBITS 45:53–59

Das D, Bandyopadhyay S (2011c) Document level emotion tagging–machine learning and resource based approach. J Comput Sist (CyS) 15(2):221–234

Das D, Anup Kumar K, Asif E, Bandyopadhyay S (2011) Temporal analysis of sentiment events-a visual realization and tracking. In: Gelbukh A (ed) Proceedings of 12th international confer- ence on intelligent text processing and computational linguistics (CICLing-2011). Lecture notes in computer science 6608, Tokyo. Springer, Heidelberg, pp 417–428

Ekman P (1992) Facial expression and emotion. Am Psychol 48(4):384–392

Esuli A, Sebastiani F (2006) SENTIWORDNET: a publicly available lexical resource for opinion mining. In: LREC'06, Genoa

Evans DK (2007) A low-resources approach to opinion analysis: machine learning and simple approaches. NTCIR, Chiyoda-ku

Fukuhara T, Nakagawa H, Nishida T (2007) Understanding sentiment of people from news articles: temporal sentiment analysis of social events. In: ICWSM'2007, Boulder, CO

Grefenstette G, Qu Y, Shanahan JG, Evans DA (2004) Coupling niche browsers and affect analysis for an opinion mining application. In: RIAO-04, Avignon, pp 186–194

Havre S, Hetzler E, Whitney P, Nowell L (2002) ThemeRiver: visualizing thematic changes in large document collections. IEEE Trans Vis Comput Graph 8(1):9–20

Hu J, Guan C, Wang M, Lin F (2006) Model of emotional holder. In: Shi Z-Z, Sadananda R (eds) PRIMA 2006. Lecture notes in computer science (LNAI) 4088. Springer, Heidelberg, pp 534–539

Izard CE (1971) The face of emotion. Appleton-Century-Crofts, New York

James W (1884) What is an emotion? Mind 9:188–205

Joachims T (1998) Text categorization with support machines: learning with many relevant features. In: European conference on machine learning, Chemnitz, 21–24 Apr 1998, pp 137–142

Kim SM, Hovy E (2006). Extracting opinions, opinion holders, and topics expressed in online news media text. In: Workshop on sentiment and subjectivity in ACL/Coling, Sydney

Kim Y, Jung Y, Myaeng S-H (2007) Identifying opinion holders in opinion text from online newspapers. In: 2007 I.E. international conference on granular computing, San Jose, CA, pp 699–702

Kobayashi N, Inui K, Matsumoto Y, Tateishi K, Fukushima T (2004) Collecting evaluative expressions for opinion extraction, IJCNLP 2004. Springer, Berlin

Kolya AK, Das D, Ekbal A, Bandyopadhyay S (2011) Identifying event–sentiment association using lexical equivalence and co-reference approaches. In: Proceedings of the workshop on relational models of semantics (RELMS 2011), ACL-HLT 19–27, Portland, OR

Ku LW, Liang YT, Chen HH (2006) Opinion extraction, summarization and tracking in news and blog corpora. In: AAAI-CAAW2006, Stanford University, CA, 27–29 Mar 2006, pp 100–107

Lafferty J, McCallum AK, Pereira F (2001) Conditional random fields: probabilistic models for segmenting and labeling sequence data. In: Proceedings of 18th international conference on machine learning, Corvallis, OR

Lin C-Y (1997) Robust automated topic identification. Faculty of the Graduate School, University of Southern California. ACL, pp 308–310

Lin KH-Y, Yang C, Chen H-H (2007) What emotions news articles trigger in their readers? SIGIR 733–734

Liu B (2009). The challenge is still the accuracy of sentiment prediction and solving the associated problems. In: 5th Annual text analytics summit, Boston, MA.

Liu B (2010) Sentiment analysis and subjectivity. In: Indurkhya N, Damerau FJ (eds) Handbook of natural language processing, 2nd edn. CRC, Boca Raton, FL

Mann WC, Thompson S (1988) Rhetorical structure theory: toward a functional theory of text organization. TEXT 8:243–281

McDougall W (1926) An introduction to social psychology. Luce, Boston

Mihalcea R, Banea C, Wiebe J (2007) Learning multilingual subjective language via cross-lingual projections. In: Annual meeting of the Association of Computational Linguistics, Prague, pp 976–983

Miller AG (1995) WordNet: a lexical database for English. Commun ACM 38(11):39–41

Mishne G, de Rijke M (2006a) Capturing global mood levels using blog posts. In: AAAI-CAAW2006, Stanford University, CA, 27–29 Mar 2006, pp 145–152

Mishne G, de Rijke M (2006b) MoodViews: tools for blog mood analysis. In: AAAI 2006 Spring symposium on computational approaches to analyzing weblogs, Stanford, CA

Mohammad S, Turney PD (2010) Emotions evoked by common words and phrases: using mechanical Turk to create an emotion lexicon. In: Proceedings of the NAACL-HLT 2010 workshop on computational approaches to analysis and generation of emotion in text, Los Angeles, CA, pp 26–34

Myers DG (2004) Theories of emotion. Psychology, 7th edn. Worth Publishers, New York, NY, p 500

Neviarouskaya A, Prendinger H, Ishizuka M (2007) Narrowing the social gap among people involved in global dialog: automatic emotion detection in blog posts. ICWSM, Boulder, CO

Ortony A, Turner TJ (1990) What's basic about basic emotions? Psychol Rev 97:315–331

Pang B, Lee L (2008) Opinion mining and sentiment analysis. Found Trends Inf Retr 2(1–2):1–135

Parrott W (2001) Emotions in social psychology. Psychology Press, Philadelphia

Plutchik R (1980) A general psychocvolutionary theory of emotion. In: Plutchik R, Kellerman H (eds) Emotion: theory, research, and experience, vol 1, Theories of emotion. Academic Press, New York, pp 3–31

Polanyi L, Zaenen A (2006) Contextual valence shifter. In: Shanahan JG, Yan Q, Wiebe J (eds) Computing attitude and affect in text: theory and applications, Chap 1. Springer, Heidelberg, pp 1–10

Popescu A, Etzioni O (2005) Extracting product features and opinions from reviews. In: Proceedings of HLT/EMNLP, Vancouver, BC, 6–8 Oct 2005

Quirk R, Greenbaum S, Leech G, Svartvik J (1985) A comprehensive grammar of the English language. Longman, London

Read J, Carroll J (2010) Annotating expressions of appraisal in English. Lang Resour Eval. doi:10.1007/s10579-010-9135-7

Seki Y (2007) Opinion holder extraction from author and authority viewpoints. In: Proceedings of the SIGIR'07, ACM 978-1-59593-597-7/07/0007

Sood S, Vasserman L (2009) ESSE: Exploring mood on the web. In: Proceedings of the 3rd international AAAI conference on weblogs and social media (ICWSM), San Jose, CA, 17–20 May 2009

Stein B, Eissen SMz (2004) Topic identification: framework and application. Paderborn University, Paderborn

Stone PJ (1966) The general inquirer. A computer approach to content analysis. MIT Press, Cambridge, MA

Stoyanov V, Cardie C (2008a) Annotating topics of opinions. In: Proceedings of LREC, Marrakech, 26 May–1 June 2008

Stoyanov V, Cardie C (2008b) Topic identification for fine-grained opinion analysis. Coling 2008: 817–824

Strapparava C, Mihalcea R (2007) SemEval-2007 Task 14: affective text. In: 45th ACL, Prague, 23–30 June 2007

Strapparava C, Valitutti A (2004) Wordnet-affect: an affective extension of wordnet. In: Proceedings of the 4th international conference on language resources and evaluation (LREC 2004), Lisbon, May 2004, pp 1083–1086

Swier RS, Stevenson S (2004) Unsupervised semantic role labelling. In: EMNLP, Barcelona

Turney PD (2002) Thumbs up or thumbs down? Semantic orientation applied to unsupervised classification of reviews. In: Proceedings of the 40th ACL, Philadelphia, pp 417–424

Voll K, Taboada M (2007) Not all words are created equal: extracting semantic orientation as a function of adjective relevance. In: Proceedings of the 20th Australian joint conference on artificial intelligence, Gold Coast, pp 337–346

Watson JB (1930) Behaviorism. University of Chicago Press, Chicago

Wiebe J, Wilson T, Cardie C (2005) Annotating expressions of opinions and emotions in language. LRE 39(2–3):165–210

Yang C, Lin KH-Y, Chen H-H (2007) Emotion classification using web blog corpora. In: Proceedings of the IEEE, WIC, ACM international conference on web intelligence, Silicon Valley, 2–5 Nov 2007, pp 275–278

Yu N (2009) Opinion detection for web content. Comput Linguistics 39

Zhang Y, Li Z, Ren F, Kuroiwa S (2008) A preliminary research of Chinese emotion classification model. IJCSNS 8(11):127–132

Discovering *Flow of Sentiment* and *Transient Behavior* of Online *Social Crowd*: An Analysis Through Social Insects

Goldina Ghosh, Soumya Banerjee, and Vasile Palade

Abstract Social media is growing at substantially faster rates, with millions of people across the globe generating, sharing and referring content on a scale apparently impossible a few years back. This has cumulated in huge participation with plenty of updates, opinions, news, blogs, comments and product reviews being constantly posted and churned in social Websites such as *Facebook, Digg and Twitter* to name a few. Even the events that are offline fetch the attention of social crowds, and considerably, their rapid sharing of views could signify the sentiment and emotional state of crowds at that particular instance. In the recent past, social media during *terrorist strikes* or natural disasters or in panic situations exhibits a tremendous impact in propagating messages among different communities and people. But the crowd participation in these interactions is grouped *on the fly,* and once the events fade out, they slowly disappear from the social media. We continuously iterate the challenges of identifying the behavioral pattern of the so-called *transient crowd* and their dispersion or convergence of sentiment and broadly answer how that could tell upon the offline events as well. While modeling the dynamics of such crowds, relevant clustering techniques have been consulted, although any method alone was not found compatible with the social media setup. The continuous cognitive pattern like a homophilic or curious and intuitive crowd with vector attributes on such social interaction motivates to incorporate an ant's or swarm's colonial behavior. Ants and swarms demonstrate well-defined chemical communication signals known as *pheromones* to segregate and distinguish specific communication patterns from cells of high concentration to those of low concentration. Hence, the *positive and negative sentiment* of transient *crowds* could be modeled, and the local influence can be measured on their posts through pheromone modeling and reinforcement of the shortest path of an ant or swarm's

G. Ghosh • S. Banerjee (✉)
Department of Computer Sciences, Birla Institute of Technology, Mesra, India
e-mail: dr.soumya@ieee.org

V. Palade
Department of Computer Science, Oxford University, Oxford OX1 3QD, UK

N. Agarwal et al. (eds.), *Online Collective Action*, Lecture Notes in Social Networks,
DOI 10.1007/978-3-7091-1340-0_3, © Springer-Verlag Wien 2014

life cycle. The primary objective of the chapter is to introduce a comparative smart methodology of ants and swarms as agent-based paradigms for investigating the community identification, namely, for *Facebook and Twitter*. The social media platforms are large enough to accommodate the ant and swarm graph for a pheromone model, tuning the time complexity of pheromone deposition and evaporation. Subsequently, the strength of association between transient users also could vary in terms of edge distribution and decay over stochastic measures of social events. We inculcate a couple of test cases fetched from *Facebook* on recent terror strikes of Mumbai, India, modeled using ants' and swarms' behavior. The results are encouraging and still in process. Empirically, the flow of sentiment and the corresponding dispersion of the crowd effect should infer or ignore a particular event, will leave a socio-computational benchmark for the mentioned proposition and will assist the ant alive in the system to reciprocate.

1 Background and Introduction

Communicating or interacting is a means of sharing ideas and placing opinion on some topic or field and also expressing different emotional feelings like being happy or expressing sadness on some events by putting some comments. In the present world of social media and networking, the means of such interaction has taken a wide range irrespective of geographical boundary, language or age. This has cumulated in huge participation, with plenty of updates, opinions, news, blogs, comments and product reviews being constantly posted and churned in social Websites such as *Facebook, Digg and Twitter* (Galuba et al. 2010) to name a few. Even the events that are offline fetch the attention of social crowds, and considerably, their rapid sharing of views could signify the sentiment and emotional state of crowds at that particular instance.

In the recent past, social media, during *terrorist strikes* or natural disasters or in panic situations exhibited a tremendous impact in propagating messages among different communities and people. But the crowds participating in these interactions are grouped *on fly,* and once the events fade out, they slowly disappear from the social media. We continuously iterate the challenge of identifying the behavioral pattern of the so-called *transient crowd* and their dispersion or convergence of sentiment (Cha et al. 2010; Kamath and Caverlee 2010) and broadly how that could tell upon the offline events as well. While modeling the dynamics of such crowd, relevant clustering techniques have been consulted, although any single method alone was not found compatible with the social media setup. The continuous cognitive pattern like a homophilic or curious and intuitive crowd with vector attributes on such social interaction motivates to incorporate an ant's or swarm's colonial behavior (Parunak 2011). Ants and swarms demonstrate well-defined chemical communication signals known as *pheromones* to segregate and distinguish specific communication patterns from cells of high concentration to those of low concentration. Hence, the *positive and negative sentiment* of transient *crowds* (Kamath and Caverlee 2011) could be modeled, and the local influence can be measured on their posts through pheromone modeling and reinforcement of the shortest path of an ant or swarm's life cycle. The primary objective of the chapter is to introduce a comparative smart methodology of ants and swarms as agent-based paradigms for investigating the community identification, namely, for *Facebook*. The social media platforms are large enough to accommodate the ant and swarm graph for pheromone models, tuning the time complexity of pheromone deposition and evaporation. We inculcate a couple of test cases fetched from *Facebook* on recent terror strikes of Mumbai, India, modeled using an ant's and swarm's behavior. The results are encouraging and still in process. Empirically, the flow of sentiment and the corresponding dispersion of the crowd effect should infer or ignore a particular event, will leave a socio-computational benchmark for the mentioned proposition and will assist the ant alive in the system to reciprocate.

If this could be the foundation of this research investigation, then the sustained implications were also envisaged with the behavior of natural ants and swarms during the conceptual layout of the proposal. Inspired by Dorigo, Ramos and others (Dorigo and Stützle 2001; Dorigo et al. 1996; Fernandes et al. 2008), the complex pattern of communication of ants with a chemical known as pheromone pointed a substantial clue to explore the temporal behavior exchange of their opinion and sentiment of crowds across a Web graph. Even such a social media and event graph also leads towards certain learning artifacts in the form of incremental learning as described by Dorigo et al. (Montes de Oca et al. 2011).

The role and reference of social media has been evenly poised from the occurrence of revolutionary social events. Considering the tune of stochastic measures of social crowd, the concept of temporal crowd and, subsequently, the strength of association between transient users also could vary in terms of edge distribution and decay over stochastic measures of social events.

The remaining part of the chapter has been organized as follows: Associated definitions and terminologies used in the context of social crowds have been elaborated in Sect. 2. Section 3 describes the prologue to understand the use of the pheromone communication model and its social modeling counterparts. Section 4 mentions certain real-life cases and proposes algorithms to address the problem. Section 4.1 discusses the results and observations from the proposal. Finally, Sect. 5 summarizes the content and mentions the scope for further research towards this direction.

2 Definitions, Terminologies and Mathematical Interpretations

Considering the structure of Facebook, it will be convenient to keep track of the post and broadcast tendency of the events through an ideal connected graph paradigm. We also consider m number of participants across social media site U, where each participant may post the messages with timestamps and will lead to a coherent campaign. Mathematically, it could be expressed as

$$M^u_i = \{m_{it_1} u_i | u_i \in U \cap m_{it_1}\} \tag{1}$$

This expression also yields concepts of forming message graphs either for strong campaigns or weak campaigns for the dispersed messages. Analytically, the content-driven campaign could be appealing when it becomes cohesive. This again can be validated if the number of edges of a subgraph for the original message graph is close to the maximal number of edges with the same number of vertices of that subgraph (Kamath and Caverlee 2011; Lee et al. 2011). There are different related definitions to conceptualize the present target model (Kamath and Caverlee 2011):

- **Transient crowd:** "A transient crowd $C \in K_t$ is a time-sensitive collection of users who form a cluster in G_t, where K_t is the set of all transient crowds in G_t. A transient crowd represents a collection of users who are actively communicating with each other at time t."
- **Time-Evolving Communication Network:** "A time-evolving communication network is an undirected graph G_t (V, E) graph with $|V| = n$ vertices and $|E| = m$ edges, where each vertex corresponds to a user in the social messaging system and an edge corresponds to a communication between two users. The weight of an edge between vertices u and v at time t is represented by $w_t(u, v)$."

3 Pheromone Communication and Social Network: Functional Analogy

Since the inception of Web 2.0, the complexity in the pattern of social interaction has been a point of investigation and emergence of the interaction pattern of social networks. The pattern of several interactions emerges from the structure of a positional reference of the person under the particular social network and the latest opinion shared by the person. Therefore, the evolution of a person-centric interest of the person for a group may be temporal and could devise the shape of the environment, resulting in a complex feedback process. Eventually, as a result of such dynamics, collective cognitive effects may emerge at the system level (across groups of people under social networks) that can influence the individuals' opinion without informing the person's. This alignment of opinions is called *consensus formation* (Parunak et al. 2011a). The coordination of exchange of opinion under social networks for a temporal event is quite similar to the feedback propagation through a shared environment known as *stigmergy,* and it can emerge a global pattern. In this chapter, we investigate such a possibility of pheromone communication envisaging social media as a container of events, and also we further analyze the temporal behavior and influence of stigmergic coordination of such events. Considering the social insect agents like ants could assign several types of pheromone in the same environment. The type of pheromone is identified by the subscripts and those assignments of pheromone that did not interact with each other. Each ant agent can drop pheromone on the ground by dropping action. Dropped pheromone gradually evaporates and diffuses in the air. Ant agents can detect diffusing pheromone only. Dropped pheromone and diffusing pheromone at position (x, y) are represented by $T_v(x, y)$ and $P_v(x, y)$ respectively (Dorigo and Stützle 2001; Fernandes et al. 2008).

$$T_v^*(x,y) = (1 - \gamma_{eva})T_v(x,y) + \sum_{k=1}^{N_a}\Delta T_v^k(x,y)$$

$$\Delta T_y^k(x,y) = \begin{cases} Q_p & \text{if } k\text{-th ant agent on the grid } (x,y) \text{ put the pheromone } v \\ 0 & \text{otherwise} \end{cases}$$

$$(2)$$

The occurrence of temporal events and pheromone evaporation initiates a stochastic probability of communication, and there is a significant convergence of opinion on social media irrespective of number of participants and group theme (Parunak et al. 2011b). It is also phenomenal that under similar theme spaces, a homogenous sample distribution under social media exhibits a reconfigurable mean and variance of space discarding the group theme at a particular instance of timestamps of the events. The proposed model also argues that in a high-dimensional social media, the attractive force between two or more participants decreases with distance and offer lower pheromone deposition and faster evaporation. The pheromone communication acts on elements that are already close to each other and defines the characteristic behavior on opinion and oral anxiety over the temporal events.

All ant agents in each case of colony are homogeneous and demonstrate the reinforcement strategy for case-specific inference on social message propagation. Each agent performs steps in the following logical chronology:

- The ant agent senses whether a food resource exists on the message graph, senses whether the message graph is a part of a nest and recognizes whether it is carrying a relevant message post under emergency.
- The ant agent might drop a certain type of pheromone depending on the output of the final termination of message. Each ant agent can use a value or type of pheromone.

- When there is positive response against the root message under emergency, if the ant agent carries no message in reply, it picks it up, and if the ant agent has a relevant support message and is on the nest part of a message graph, it drops it.
- Even the sense of direction can also be an indication for the implication of the final transient mood of the crowd.

The relevant application also supports the present study to identify the potential link of Facebook group participation with viral advertising responses. The results suggest that college-aged Facebook group members are generally involved in higher levels of self-disclosure and maintain more favorable attitudes toward social media and advertising compared to non-group members (Chu 2011). Similarly, fundraising events for a cause also deployed a Facebook campaign and received a substantial impact of opinion diffusion and similarity toward a specific social call (Kamath and Caverlee 2010).

The inclusion of pheromone communication creates deliberate space with the concept of transient crowd in social media (Kamath and Caverlee 2011). Transient crowds are dynamically formed and have a short span of life. We interpret and explore the stochastic relationship of time-evolving graphs for transient and temporal crowd formation on Web media. The participants of these social networks may be clustered along a number of dimensions including content-based or thematic interest or may be diversified geographic locations driven toward the same interest. This concept motivates us to incorporate the concept of dynamic clustering of the time-evolved graph. In this particular proposed model, the structure of edges is changing. The interesting relationship between transient crowds for a particular time instant and the swarm's behavior has already been identified. The dense coverage of edges is prone to demonstrate the distinguished clusters shown in several contemporary literatures (Saha and Mitra 2006). The proposed model inculcates a flow of sentiment and opinion over a post analytically, and we also solicit certain contextual definitions to point out the foundation of the proposal elaborated in Sect. 2 (Kamath and Caverlee 2011).

4 Presentation of Data Snippets and Analysis with Proposed Model

The social network sites could be contemplated as a temporal media for transferring crucial events and drawing the attention of different people. This is usually done (Burke et al. 2010; Leskovec et al. 2008) when there is a requirement of sudden critical social causes. The evidence is obtained from Facebook. The recent *Mumbai blast* had experienced casualties in large scale, and many of them required blood. This issue became a crisis, and one of the common citizens posted some photographs on his Facebook wall of the blood donation camp in order to seek help for the sake of those affected in the blast. Seeing this post, as many as 14 people started communicating with him immediately on Facebook either through the comment

Table 1 Temporal event propagation under Facebook

Name of the FaceBook participants	Date	Time stamp	Information shared
Mohit Sharma	July 14, 2011	7:37 p.m.	i have also denated blood on this camp
Rajendra Suryawanshi	July 14, 2011	11:21 a.m.	Go to Mumbai
Shreekant Jagtap	July 14, 2011	11:59 a.m.	Blood group B+
Vaishnavi-piyali Suratwala	July 14, 2011	12:58 p.m.	AB+
Prashant Suratwala	July 14, 2011	1:21 p.m.	B+
Pinkesh Suratwala	July 14, 2011	4:29 p.m.	B+
Manish Suratwala	July 14, 2011	5:15 p.m.	my blood group is b+ve and i am willing to go to Mumbai
Milind Joshi	July 14, 2011	6:24 p.m.	O+
Ranjit Bansode	July 14, 2011	8:19 p.m.	I m also with u.My bld grp is A+ n my wife's O+
Dhawal Manojkumar Suratwala	July 14, 2011	11:02 p.m.	if u need i will surely cum to help u out
Milind Joshi	July 15, 2011	6:03 p.m.	good job man
Chandan Shantilal Suratwala	July 15, 2011	9:32 p.m.	thanks to u all to support my appeal...we will surely do our best to make our pune best
Viraj Gelada	July 16, 2011	7:12 p.m.	my blood group b+...
Dhawal Manojkumar Suratwala	July 16, 2011	10:43 p.m.	will always support u and even pune 4 ne kind of work.

box or through short messages in their phones on the same date and on the following dates. The information obtained from the wall post has been described in Table 1 (retrieved from http://www.facebook.com):

It is also noticed in Cho et al. (2011) that the information was transferred to different parts of the country at different time instants to different individuals through other groups or people via an obvious interconnected fashion. The formations of connected networks and cascaded events are deliberate, and they also exhibit a structure of graph in their representation. Table 2 demonstrates the evidence of the said event.

4.1 Proposed Algorithm and Analysis

As Table 2 extracts the snippet that how diversified people, irrespective of regions, could have been accumulated in shared environments for sharing their opinions (Cho et al. 2011; De Choudhury et al. 2010). Considering these two attributes of

Table 2 Dataset of Facebook based on the Mumbai blast

Name of the event	Personal id number	Location	Start time	End time
Support the people of Mumbai in blast	246599782018044	Zaveri Bazar OPERA HOUSE	2012-07-13T13:30:00	2012-08-15T15:30:00
Need 10,00,000 Supporter to protest Against Blast in Mumbai-Invite 100+	240232295996771	http://www.facebook.com/imanojsingh	2012-01-26T01:00:00	2012-01-26T01:00:00
Again blast in Mumbai	161485890589611	Jaipur, Rajasthan	2012-07-14T00:30:00	2012-07-14T03:30:00
Again blast in Mumbai	247294775280525	Jaipur, Rajasthan	2012-07-14T00:30:00	2012-07-14T03:30:00
Again blast in Mumbai	219309511444604	Jaipur, Rajasthan	2012-07-14T00:30:00	2012-07-14T03:30:00
To see bomb blast in Mumbai	264469923566556	Mumbai	2013-07-13T07:00:00	2013-07-13T09:00:00

record sets, a schema can be conceptualized as shown to present the distribution of data for different perspectives (Scheme 1).

Computationally, we can generate the rate of flow of information among different individuals denoted as nodes, and the links between these nodes are known as edges. This is analogous to the proposed pheromone communication to indicate reinforcement of a particular edge as per the pheromone deposition and evaporation rule followed by natural insects (Dorigo and Stützle 2001).

A node takes the initiative to send a message to its connected links. These linked nodes can again spread this message to all other connected nodes. In Fig. 1, it is seen that Node 1 is the original sender of information. Node 2 and Node 4 are the receivers. Node 3 and Node 5 receive the message from Node 2 and Node 4 respectively. Thus, if there are other links between Node 2, 3, 4, or 5, then they will also get the message through them. As preprocessing steps, we incorporate MATLAB to simulate the interaction diagram for the transfer of messages and find the difference in characteristics of path that in turn provide a substantial insight to further analyze its semantics.

The rate of message transformation will depend on the number of connectivity each node has of itself, that is, the rate of flow of messages is

$$R_i = n^m \tag{3}$$

where R_i is the rate of information flow $R_i = \{r_1, r_2, \ldots r_t\}$, n is the number of nodes that received the message first $n_i = \{n_1, n_2, \ldots n_t\}$, m is the number of links in the node $= \{1, 2, \ldots t\}$.

Figure 2 shows the rate of flow of information among the other nodes with the direction of the flow of messages.

Scheme 1 Distribution of sharing

Fig. 1 Propagation of message from the main sender

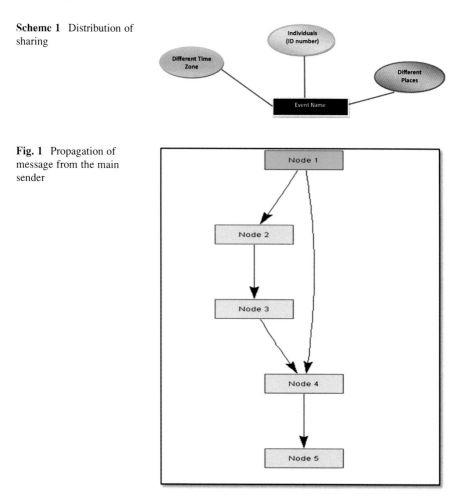

Here, Node 1 is left isolated since it has already sent the message to its connected links. It is the responsibility of other connected nodes to transfer the message. In the figure, Node 2 is sending the message to Node 6, and Node 3 is sending it to Node 7. Similarly, Node 4 is sending it to Node 9, and Node 9 delivers it to Node 5 and Node 8. With the help of this propagation style, the message or some events are spread through a social network through connected links. Node 1, being the original sender, will always wait for either some response or some positive effect from the receivers. In Fig. 3, it is shown how all the nodes ultimately got connected to the information given by Node 1 and even shared opinion or transferred the message between themselves and Node 1.

Immediately before the event of the Mumbai bomb blast, the particular participant had general discussions among his community in Facebook. This information that is shared in the communication was very conventional and of lesser social message value, thus as likely as blood donating camp. A survey report on this issue

Fig. 2 Transmission of the
message via other
connected links in a social
network

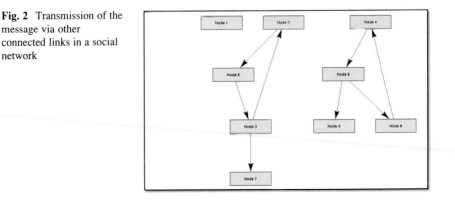

Fig. 3 Flow of message to
all connected link leading to
communicate with the
primary sender of the
message

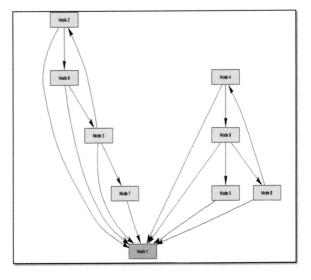

is given in Table 3. As seen, there are no crucial messages; hence, there will be no
option of spreading them in a wide range. Thus, the rate of flow of the information
will also depend on two factors: Firstly, it will depend on the weightage of the
message, that is, how important the message is. If it is a vital issue, then it will be
propagated to all the different connective nodes. If not, then the flow will be
restricted. Secondly, if the information is forwarded, then also the response rate
will be much lower.

In Fig. 4, the description of the message flow exists where only Node 2 transfers
the information to Node 7 and Node 5 transfers it to Node 8, Nodes 3, 4, 6, and 9 are
isolated in the graph since they did not transfer the message, whereas it is seen that
Node 3 had transferred the message to Node 7 in Fig. 2. With respect to Fig. 1, it is
also seen that there is an initial connectivity between Node 1 and Node 2 and Node
1 and Node 4, which means that the information is passed to Node 2 and Node
4. Node 2 transfers the message to Node 3 and Node 4 to Node 5. Still, in Fig. 3, we

Table 3 Data snap from Facebook of the root initiator immediate before *Mumbai Blast* [https://www.facebook.com/chandan.suratwala (Refer Mr. Chandan Shantilal Suratwal)]

Types of post	Date	Time	Responses
Shared information of donating eye in H.V.DESAI EYE HOSPITAL	July 13, 2011	11:35 a.m.	3 people like this
A general broad cast of a poetry on life	July 12, 2011	12:28 a.m.	2 people like this.
Uploaded 5 photos and using the tag line "MY PUNE'S WORST BLACK DAY 12TH JULY 1961"	July 11, 2011	10:35 p.m.	No comments made
Share an opinion on a simple matter	July 11, 2011	12:33 p.m.	6 people like this.
Made a comment on a tourist travel	July 11, 2011	10:21 p.m.	No moments made

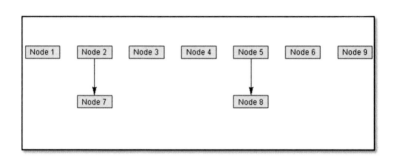

Fig. 4 Selected nodes as recipients

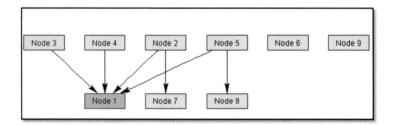

Fig. 5 Node 1 being the originator of message obtains response from different nodes

observe that Nodes 3 and 4 do not take the initiative to transfer the information further to the other nodes.

As Node 1 was the original sender of the information, it will again wait for some response or reaction from other nodes. But since the message or event is not very important, so the propagation of information will be less compared to the previous case, and also the response might or might not occur. In Fig. 5, we observe that Nodes 2, 3, 4, and 5 only respond to the message that Node 1 sent, whereas Node 7 and Node 8, although receiving the message, did not respond. Compared to Figs. 3 and 5, it has much less connectivity or interactions among the nodes.

4.2   Validating the Flow of Information

The information that is obtained from Table 1 gives a detailed explanation about the date, time, and some vital information related to the event of the Mumbai blast and where a blood donation camp required blood. Similarly, from Table 3, we obtain the information in respect to some basic common discussions before the event of the Mumbai blast. These messages are obtained from "Mr. Chandan Shantilal Suratwal's" Facebook account. Now, we can create a survey report based on reference (Backstrom and Leskovec 2011) that will help in testing the *"Rate of Sentiment Flow."* Firstly, let *"u"* be the number of his friends who receive the message. Then, if those friends find the information to be important and feel that it needs to be shared, then they will be transferring it to more of his friends. In this way, the flow of messages will take place. Gradual transfer of the messages can take place in *"n"* levels of propagation. The initial set of friends who had obtained the message first from the sender always remains fixed, say, x_p. Therefore, the increase in members for the flow of information at each level is given by $(1 + X_p/n)^n$. We consider two parameters, that is, "the date of propagation" and "time in Hours," which is denoted as $D_T(H)$.

The rate of sentiment flow is inversely proportional to time $D_T(H)$.

The rate of sentiment flow is proportional to the number of people who received the message. Therefore, we can frame that

$$\text{Rate of sentiment flow} = k \times \frac{\text{Number of People}}{D_T(H)}$$

The final Flow Rate of Sentiment $F(S)_r$

$$= \frac{1}{\exp(D_T(H))} \left[X_p \times \sum_{n=1}^{\infty} \left(1 + \frac{X_p}{n} \right)^n \right]. \tag{4}$$

Now, by considering the status of Table 1, we design the "Rate of Flow of Sentiment" among different friends and their friends of friends in a different date and time. Since an event like the Mumbai blast is a very vital and a sensitive issue, hence this information needs to flow at a very fast rate to many people within a short period of time.

The event was uploaded on June 14 at 11 a.m.; soon after that, the propagation of messages starts taking place. Initially, only four of the friends receive the message. They forward it to their friends, and then further those friends send it to their friends of friends. Here, totally 11 propagations take place. At each level, there is an increase of receivers. Table 4 gives the complete picture of this scenario, which is actually based on Eq. (2). The rate of flow of information increases exponentially as time increases soon after the uploading of events takes place, but the flow rate will decrease as the number of days increases. This is so since the event is important and sensitive, so the delay in propagation is not appreciated. This is proved in Figs. 6 and 7.

Here, in Fig. 6 we see that there is a gradual increase in the flow of information on June 14 as time duration increased. There is an exponential curve that denotes the increase in the flow of sentiment. As time passes, the value of the message

Table 4 Evaluation table for detecting the flow rate of sentiment

Date of propagation	Propagation time (Hrs) $D_T(H)$	1/exp $(D_T(H))$	First level receivers (X_p)	Number of propagation (n)	$1 + (X_p/n)^n$	Flow rate of sentiment $(F(S)_r)$
July 14, 2011	11:21 a.m.	0.06	4	2	9	2.16
July 14, 2011	11:59 a.m.	0.31	4	3	12.7	3.93
July 14, 2011	12:58 p.m.	3.21	4	4	16	205.4
July 14, 2011	4:29 p.m.	41.3	4	5	18.8	3121.5
July 14, 2011	6:24 p.m.	7.4	4	7	23.6	22750.4
July 14, 2011	7:37 p.m.	8.61	4	1	5	172.3
July 14, 2011	8:19 p.m.	9.31	4	8	25.62	143403.33
July 15, 2011	6:03 p.m.	27.77	4	9	27.57	3040.25
July 16, 2011	7:12 p.m.	3895.5	4	10	28.92	45057.36
July 16, 2011	10:43 p.m.	3.61	4	11	30.31	1.14

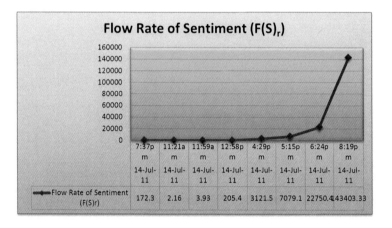

Fig. 6 Flow rate of sentiment on 14 June

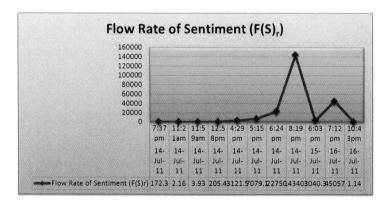

Fig. 7 Flow rate of sentiment from 14 June to 16 June

gradually decreases since the requirement gets fulfilled within a particular time limit. Thus, the curve gradually decreases from June 15 and 16. This phenomenon is shown in Fig. 7.

High Level Description of the *Pheromone Crowd* Algorithm

```
/* Parameter list: Message graph of FaceBook,, Social
crowd Set (P_si ), Message with time instant (m_it₁)
 Pheromone evaporation and dropping as per equation (1),
'
set of colour paths red, green and blue on message graph
*/

Let   P_si is the social crowd set for immediate time
instant and P_si-1 is the immediate predecessor node

for every crowd  P_si and for every message m,  P_si
```

$$\in P_s m_{it_1} u_i \text{ do}$$

```
        if  m_it₁  is  a recent message under t time
instant

        Then create a green edge on message graph and
call:
```

$$T_v^*(x,y) = (1 - \gamma_{eva})T_v(x,y) + \sum_{k=1}^{N_a} \Delta T_v^k(x,y)$$

```
        else

Get the parent current set of recipient message as M ∈ u
with maximum similar messages. If   there is more than
one participants with same number of common nodes
connected or shared then select P_si

Create a new blue continuous edge and insert a directed
edge with the immediate previous node from the root
initiator to  P_si

            else

    Insert red edge for indirect connection with other
nodes of same message graph

                            end if

        end for

    Traverse entire block of message graph, that does not
have any child node

            do
                Mark the pheromone with green, red or
blue

        end for
```

4.3 Post-Simulation Experience and Visualization

After initial modeling on data sets acquired from Facebook snaps, the temporal rating of event deceleration (e.g., blood donation request for casualties) for a given time instant has been visualized through the *Python library standard* with a standard hardware setup. As shown in Fig. 8, the blocks of the posts are shown; the red indication implies the connectivity with other nodes that are indirectly connected with the root request. Absent pheromones signify if and only if there is no substantial response of the request within the intra and inter nodes as well. Presence of pheromone describes the reinforcement of message requests, and hence the entropy seems more explicit with oral anxiety enhancements in the social group. The timestamp rating has been simulated from 5 units to 20 units, and anxiety on opinion also becomes slightly enlarged. It cannot be inferred that oral anxiety is a function of the duration of temporal events and the size of transient crowds, but the pheromone map shown in red and green polygons could be able to define it. The bubbles shown are the nodes of the social group where the temporal events take place with the participants of crowds. There may be certain participants who are transient in nature in this event.

Continuing the analysis of the pheromone-based model for sentiment flow, the model incorporates a Z score. The most general way to obtain a Z score is to accomplish a Z test. This is to define numerical test statistics that can be calculated from a collection of data, such that the sampling distribution of the statistic is approximately normal under the null hypothesis. Statistics that are averages (or approximate averages) of approximately independent data values are generally well approximated by a normal distribution. An example of a statistic that would not be well approximated by a normal distribution would be an extreme value such as the sample maximum.

The standard score is

$$z = \frac{x - \mu}{\sigma} \tag{5}$$

where x is a raw score to be standardized, μ is the mean of the population, and σ is the standard deviation of the population.

The quantity z represents the distance between the raw score for the responses made against the social post and the population, meaning the total number of Facebook participants in units of the standard deviation. Z is negative when the raw score is below the mean, positive when it is placed above. The red line indicates Z scores against the normal Facebook posts as shown in Figs. 4 and 5, whereas the blue line indicates the presence of pheromone-based reinforcement under the message posts in emergency. Figure 9 puts an analysis for the absence ratings of responses from 6 units' time instants to 14 units since this span of time is given for recording the responses. We concentrate on the Z score due to its population mean and population deviation for messages and participants respectively of the social network (Fig. 10).

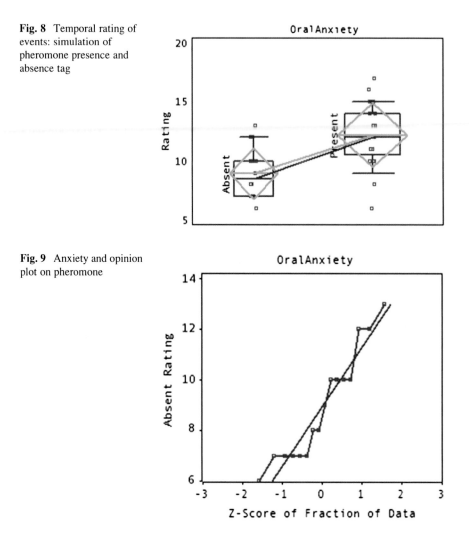

Fig. 8 Temporal rating of events: simulation of pheromone presence and absence tag

Fig. 9 Anxiety and opinion plot on pheromone

The probability distribution for each set of different opinions against the message post comparing two probability distributions can be obtained by plotting their quantiles against each other. First, the set of intervals for the quantiles are chosen. A point (x,y) on the plot corresponds to one of the quantiles of the second distribution (y-coordinate) plotted against the same quantile of the first distribution (x-coordinate). Thus, the line is a parametric curve with the parameter, which is the (number of the) interval for the quantile. As the difference of opinion is explicit in responses, therefore a typical cumulative distribution of opinion and flow could be plotted. Here also, the blue line indicates the quantile plot of responses under emergency.

Finally, a box plot of normalized data has been presented demonstrating the message continuity over the Facebook message graph. There are instances of pheromone dispersion and discontinuity in post simulation, and the *reliability* of

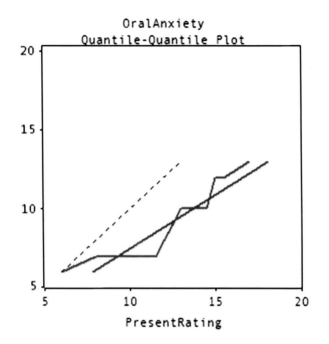

Fig. 10 Dispersion of opinion on different nodes

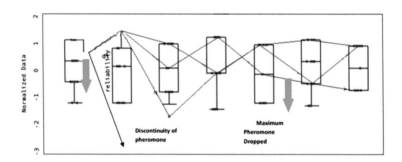

Fig. 11 Community response and pheromone plot

the communication was visible during emergency by certain groups. The base line of the plot in Fig. 11 follows the univariate range from 5 to 25 units sessions on which the transient crowd accumulated and dispersed, although the pheromone threshold value has been threshold as random and as per the dataset collected it assigns approximate maximum pheromone dropped became maximum with a few box plot only on the message graph. The plot is still an estimation with one case study, and more accuracy could be devised if a few similar instances of crisis responses of Facebook could be collected.

From all these observations, we emphasize on statistical simulation derived from pheromone assignment. XML extraction of the semantic relation of each post may reveal extended versions of transient behavior. As also shown, the immediate previous simulation (Figs. 4 and 5) of the same message graph before the crisis also concentrates on nodal analysis without pheromone population, but nontransmission of information and isolation of nodes were also clearly evident leading toward expected temporal tendency of social crowds.

5   Conclusion and Further Scope of Research

In this chapter, we have investigated the metaphorical relationship of a swarm's pheromone map with the sentiment and opinion flow of transient crowds of social media under particular situations of crisis. We present a case study of such crowds and message boards with opinion flows from Facebook, and the same message graph is referred to distinguish the crisis and precrisis paradigm. Analytically, we present a novel pheromone-driven algorithm to trace such events and flows of sentiment of the crowd accumulated for a particular theme on social media. The preprocessing and post-simulation experiments depict interesting observations of transient crowds and their opinion propagation. Pheromone tracing has been proposed as a compatible and justified tool for such stochastic and time-bound social graph analysis scenarios. The model can be well placed under the analysis of tweets, although certain other hybrid optimization algorithms, e.g., clustering, could also be incorporated. From a technical implementation point of view, XML semantics and nodal analysis could reveal empirical validation as more realistic. Interfacing the semantic analysis of XML and MATLAB simulation would be a good challenge if more Facebook instances could have been collected.

As part of our future work, we plan to develop a hybrid algorithm from these experiments to further explore social graph mining perspectives. We also plan to investigate hybrid social graph clustering approaches for implementation.

References

Backstrom L, Leskovec J (2011) Supervised random walks: predicting and recommending links in social networks. In: WSDM'11, 9–12 Feb 2011, Hong Kong, China
Burke M, Marlow C, Lento T (2010) Social network activity and social well-being. In: CHI 2010, Atlanta, GA, 10–15 Apr 2010
Cha M, Haddadi H, Benevenuto F, Gummadi KP (2010) Measuring user influence in Twitter: The million follower fallacy. In: Fourth international AAAI conference on weblogs and social media, Washington, DC, 23–26 May 2010
Cho E, Myers SA, Leskovec J (2011) Friendship and mobility: user movement in location-based social networks. In: KDD'11, San Diego, CA, 21–24 Aug 2011

Chu S-C (2011) Viral advertising in social media: participation in Facebook groups and responses among college-aged users. J Interact Advert 12(1, Fall):30–43

De Choudhury M, Mason WA, Hofman JM, Watts DJ (2010) Inferring relevant social networks from interpersonal communication. In: WWW 2010, Raleigh, NC, 26–30 Apr 2010. ACM, 978-1-60558-799-8/10/04

Dorigo M, Stützle T (2001) An experimental study of the simple ant colony optimization algorithm. In: Mastorakis N (ed) Advances in fuzzy systems and evolutionary computation, Artificial intelligence series. World Scientific and Engineering Society Press, Dallas, TX, pp 253–258

Dorigo M, Maniezzo V, Colorni A (1996) Ant system: optimization by a colony of cooperating agents. IEEE Trans Syst Man Cybern B 26(1):29–41

Fernandes C, Merelo JJ, Ramos V, Rosa A (2008) A self-organized criticality mutation operator for dynamic optimization problems. In: Keijzer M (ed) Proceedings of GECCO'08 - 10th annual conference on genetic and evolutionary computation. ACM Press, Atlanta, GA, pp 937–944

Galuba W, Chakraborty D, Aberer K, Despotovic Z, Kellerer W (2010) Outtweeting the Twitterers – predicting information cascades in microblogs. In: 3rd Workshop on online social networks (WOSN 2010), Boston, MA, 22 June 2010

Kamath KY, Caverlee J (2010) Identifying hotspots on the real-time web (Short paper). In: Proceedings of 19th ACM international conference on information and knowledge management (CIKM 2010), Toronto, ON

Kamath KY, Caverlee J (2011) Transient crowd discovery on the real-time social web. In: Proceedings of 4th ACM international conference on web search and data mining (WSDM 2011), Hong Kong, China

Lee K, Caverlee J, Cheng Z, Sui DZ (2011) Content-driven detection of campaigns in social media (short paper). In: 20th ACM international conference on information and knowledge management (CIKM), Glasgow

Leskovec J, Backstrom L, Ravi Kumar, Tomkins A (2008) Microscopic evolution of social networks. In: KDD'08, 24–27 Aug 2008, Las Vegas, NV

Montes de Oca M, Stützle T, Van den Enden K, Dorigo M (2011) Incremental social learning in particle swarms. IEEE Trans Syst Man Cybern B 41(2):368–384

Parunak HVD (2011) Swarming on symbolic structures: guiding self-organizing search with domain knowledge. In: Proceedings of the eighth international conference on information technology: new generations (ITNG 2011), Las Vegas, NV. IEEE, Piscataway, NJ

Parunak HVD, Brueckner SA, Downs E, Yinger A (2011) Opinion dynamics with social constraints and exogenous drivers. In: Cultural and opinion dynamics workshop at ECCS 2011 (CODYM 2011), Vienna

Parunak HVD, Downs E, Yinger A (2011) Socially-constrained exogenously-driven opinion dynamics. In: Fifth international IEEE conference self-adaptive and self-organizing systems (SASO 2011), Ann Arbor, MI

Saha B, Mitra P (2006) Dynamic algorithm for graph clustering using minimum cut tree. In: ICDMW '06, Washington, DC, USA. IEEE Computer Society, Piscataway, NJ, pp 667–671

Collective Emotions Online

Anna Chmiel, Julian Sienkiewicz, Georgios Paltoglou, Kevan Buckley,
Marcin Skowron, Mike Thelwall, Arvid Kappas, and Janusz A. Hołyst

Abstract This chapter analyzes patterns in messages posted to several Internet
discussion forums from the perspective of the sentiment expressed in them and the
collective character of observed emotions. A large set of records describing comments expressed in diverse cyber communities—blogs, forums, IRC channels, and
the Digg community—was collected, and sentiment classifiers were used to estimate the emotional valence (positive, negative, or neutral) of each message. A
comparison with simple models showed that the data included clusters of comments
with the same emotional valence that were much longer than similar clusters
created by a random process. This shows that there are emotional interactions
between participants so that future posts tend to have the same valence as previous
posts. Threads starting from a larger number of negative comments also last longer
so negative emotions can be treated as a kind of discussion fuel; when the fuel
(negativity) is used up in the discussion, it may finish. Moreover, the amount of user
activity in a particular thread correlates positively with the presence of negative
emotions expressed by the individual user in the thread. In summary, the analyses
describe individual and collective patterns of emotional activities of Web forum
users and suggest that negativity is needed to fuel important discussions.

A. Chmiel • J. Sienkiewicz • J.A. Hołyst (✉)
Center of Excellence for Complex Systems Research, Faculty of Physics, Warsaw University
of Technology, Koszykowa 75, 00-662 Warsaw, Poland
e-mail: jholyst@if.pw.edu.pl

G. Paltoglou • K. Buckley • M. Thelwall
Statistical Cybermetrics Group, School of Technology, University of Wolverhampton,
Wulfruna Street, Wolverhampton WV1 1LY, UK

M. Skowron
Interaction Technologies Group, Austrian Research Institute for Artificial Intelligence,
Freyung 6/3/1a, 1010 Vienna, Austria

A. Kappas
School of Humanities and Social Sciences, Jacobs University Bremen, Campus Ring 1, 28759
Bremen, Germany

N. Agarwal et al. (eds.), *Online Collective Action*, Lecture Notes in Social Networks,   59
DOI 10.1007/978-3-7091-1340-0_4, © Springer-Verlag Wien 2014

1 Introduction

The notion of *collective action* is usually associated with the responses of group members taken in order to maintain or improve the group's conditions (Wright et al. 1990). Research in this area frequently touches on the issue of *prediction* of the internal conditions that have to be fulfilled in order for the action to be undertaken [e.g., moral convictions (van Zomeren et al. 2012), injustice, efficacy, identity (van Zomeren et al. 2008)]. However, it seems that there is also a key component—emotions—that can play a pivotal role as an inhibitor of collective action, both directly (Stürner and Simon 2004) and indirectly (Taylor 1995). Moreover, the affective component can lead to the emergence or suppression of collective behavior (Sabucedo et al. 2011).

The Internet can be treated as a system of human behavior in which social dynamics are evident and measurable (Onnela and Reed-Tsochas 2010; Barabási 2005; Huberman et al. 1998; Sobkowicz and Sobkowicz 2010; Mitrović et al. 2010; Szell et al. 2010; Castellano et al. 2009; Kujawski et al. 2007; Chmiel et al. 2009). Communication in this medium displays different activity patterns compared to traditional communication (Radicchi 2009). People increasingly spend time using sites like MySpace, Facebook, Twitter, and blogs, and hence e-communities (Walther and Parks 2002) have become widespread and important. It is also obvious that collective actions take place in the Internet (Postmes and Brunsting 2002)—this medium can serve as both creator and transmitter of this notion [as in the case of Spanish "15M movement" (Borge-Holthoefer et al. 2012)]. The analysis of emotional interactions in e-communities is crucial for obtaining a comprehensive insight into social relations. In this study, we focus on collective emotional behavior, modeling its emergence, and the conditions that are necessary for it to happen.

Although emotions are typically expressed using a variety of nonlinguistic mechanisms—such as laughing, smiling, vocal intonation, and facial expression—textual communication can be just as rich and can be augmented by expressive textual methods—such as emoticons and slang (Gamon et al. 2005). Taking advantage of this, sentiment analysis, a research field in computational linguistics and computer science, has evolved rapidly in the last 10 years in response to a growing recognition of the importance of emotions in business and the increasing availability of masses of text in the social Web. The development of a number of algorithms to detect positive and negative sentiment has also made large-scale online text sentiment research possible, such as predicting elections by analyzing sentiment in Twitter (Tumasjan et al. 2010) and diagnosing trends for happiness in society via blogs (Dodds and Danforth 2010) and Facebook status updates (Pang and Lee 2008).

In this chapter, we discuss the impact of emotional expressions of Internet users on the vitality of online debates. We focus on (1) measuring the transfer of emotions between participants, (2) the influence of emotions on a thread's life span, and (3) the relationship between user behaviors and the emotionality of a discussion.

We especially focus on manifestations of collective emotional behavior in Sect. 3 while comparing emotional distributions with random equivalents. Section 4 touches on another interesting aspect of emotionally driven online discussions—the conditions that have to be fulfilled in order for a discussion to be fruitful (measured in the number of utterances). Finally, the BBC Forum analysis in Sect. 5 investigates the sentiments expressed by individual users within discussions at both global and local levels.

2   Data

We collected over six million comments from four prominent different interactive spaces: blogs, BBC discussion forums, popular social news Web site Digg, and #ubuntu IRC discussion channel. The texts were processed using sentiment analysis classifiers to predict their emotional valence (see Sect. 2.2), which according to a converging evidence is at the center of emotion experience and is considered one of the most important determinates of behavior from simple life-forms to humans (Lang and Davis 2006).

2.1   Datasets

The British Broadcasting Corporation (BBC)[1] (Chmiel et al. 2011a) Web site has a number of publicly open *Message Boards* covering a wide variety of topics that allow registered users to start their own discussions and post comments on existing discussions. Comments are postmoderated and anything that breaks the House Rules is subject to deletion. Our data included discussions posted on the Religion and Ethics and World/UK News message boards starting from the launch of the Web site (July 2005 and June 2005 respectively) until the beginning of the crawl (June 2009)—these were found to have interesting emotional content. The dataset comprises 100,000 discussions, 2.5 million comments and 18,000 users.

The Blogs dataset is a subset of the Blogs06 (Macdonald and Ounis 2006; Weroński et al. 2012) collection, which is an uncompressed 148 GB crawl of approximately 100,000 different blogs (more than three million Web pages) and spans 11 weeks, from 6 December 2005 to 21 February 2006. The subset was created by manually removing the HTML of the Blogs06 collection and keeping only the actual content (i.e., posts and comments). Only posts attracting more than 100 comments were extracted, as these seemed to initialize nontrivial discussions.

The Digg dataset comprises a full crawl of digg.com, one of the most popular social news Web sites. The analysis spans February to April 2009 and consists of all the stories, comments, and users that contributed to the site during this period.

---

[1] http://www.bbc.co.uk

The resulting dataset contains approximately 1.9 million stories, 1.6 million comments, and 800,000 users (Paltoglou et al. 2010; Pohorecki et al. 2013).

The Internet Relay Chat (IRC) is a medium that allows maintaining real-time multiuser discussions. The presented dataset contains information from the logs of #ubuntu discussion channels dating between 1 January 2007 and 31 December 2009. Data were preprocessed and transformed into a structure of over 90,000 one-to-one dialogues with almost 1.9 million comments (Sienkiewicz et al. 2013).

2.2 Algorithms

Sentiment analysis algorithms typically operate in three stages: (a) separate objective from subjective texts, (b) predict the polarity of the subjective texts, and (c) detect the sentiment target (Paltoglou et al. 2010). A variety of methods are used, including machine learning based upon the words used in each text, summarized in vector form (Riloff et al. 2006), and lexical approaches that start with a dictionary of known sentiment-bearing terms and apply linguistically derived heuristics to predict polarity from their occurrence and contexts (Wilson et al. 2009). One ongoing problem, however, is domain transfer—algorithms need to be tailored for each type of text that they are applied to because words tend to have different meanings in different contexts. In consequence, large sets of human-annotated data are needed to train and evaluate systems for each new application domain.

The algorithm that we used to detect and characterize the emotional content of posts is based on standard, supervised machine-learning principles (Sebastiani 2002). During a *training* phase, a corpus of documents is provided to the algorithm, i.e., a set of documents each one belonging to a specific category. The category of each document was determined during a preceding *corpus development* process by human experts who read its content and manually classified it. The algorithm extracts the characteristics of each class by analyzing the provided documents, i.e., "learns by example," and stores this knowledge. Subsequently, during the *application/testing* phase, the algorithm applies the acquired knowledge to new, unseen documents and determines the best category under which they can be classified.

In this study, we implemented a hierarchical extension of a standard Language Model classifier (Sebastiani 2002) (abbreviated as h-LM). LM classifiers are a typical example of probabilistic classifiers, which estimate the probability that a document belongs to all of the available classes and select the one with the highest probability as the final prediction. In our hierarchical extension, a document is initially classified by the algorithm as objective or subjective, and in the latter case a second-stage classification determines its polarity, either positive or negative.

We used a manually annotated subset of Blogs06 dataset as *training corpus*. Human assessors examined approximately 34,000 documents for whether they contain factual information or positive/negative opinions about specific entities, such as people, companies, and films, and assigned a category to each document.

Because the distribution of documents per category is uneven in the specific corpus, the probability thresholds for both classification tasks were optimized on a small subset of humanly annotated BBC comments. The optimized classifier has an average F1-value of 66.68 % on the subjectivity detection task and 60.93 % on the polarity detection task on a humanly annotated BBC subset.

3   Cluster Distribution

To detect affective interactions between discussion participants, we calculated statistics for groups of comments with similar emotion levels. Each thread was transformed into a chain structure even if the the original structure was tree-like structure was present (see Fig. 1). The four datasets on which sentiment analysis was performed are characterized by different structures. In the case of Blogs and IRC dialogues, all posts were originally arranged in chains of successive comments, as shown in Fig. 1b. In other words, new posts are automatically added after the last post. On the other hand, the BBC and Digg data are arranged in a forum-like structure (see Fig. 1a), meaning that each user may make a comment to any previous post, thus starting a separate discussion. However, for the purpose of this study, each discussion (thread) in each dataset was arranged chronologically, as in Fig. 1b. In this way, it was possible to compare these different communities. Although BBC and Digg have a forum-like structure, the default view as presented to the user was chronological. Thus, a chronological simplification for analysis can be justified.

   We define an emotional cluster of size n as a chain of n consecutive messages with the same sentiment orientation, i.e., negative, positive, or neutral, where before the cluster and after it there is message with the valence different from the cluster valence (see upper row of Fig. 2) (Chmiel et al. 2011b). For the purpose of comparison, we show also the shuffled data obtained from the same discussion (see bottom row of Fig. 2), which display clearly shorter clusters than those in the original data. A possible explanation of the observed fact could be the emotional interactions between the participants of the discussion in the original data.

   In order to test this hypothesis, we checked the complementary cumulative distribution function (CCDF) $P^{(e)}(\geq n)$ that describes the frequency of clusters of size greater than or equal to n for the cases of negative, positive, and neutral emotions, $e = \{-1, 0, 1\}$. The results are presented as symbols in Fig. 3. For comparison, the corresponding CCDF from the independent and identically distributed (i.i.d.) random process

$$P^{(e)}_{\mathrm{iid}}(\geq n) = p(e)^{n-1} \qquad (1)$$

(dotted lines) is also plotted, where $p(e)$ is the probability of negative, positive, or neutral emotion (see Table 1). The i.i.d. random process (Feller 1968) is an effect

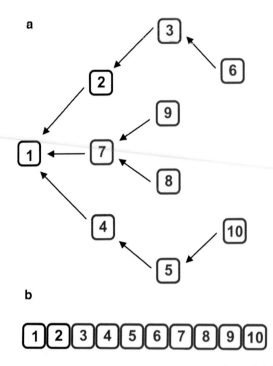

Fig. 1 The difference between the actual tree structure (**a**) present in the BBC and Digg datasets as compared to the chronological layout of the posts (**b**), e.g., originally present in Blogs and IRC. The *numbers* indicate the order of messages (1 being the first, 10 being the last) while *arrows* indicate that a post was given in reply to another one (e.g., post 9 is the response to post 7)

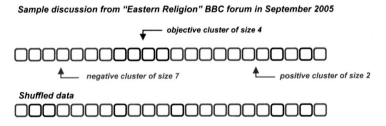

Fig. 2 An example of a discussion in the "Eastern religion" BBC Forum. The original thread that consists of 22 posts is shown in the *upper row*. Each *box* represents one post. *Red*, *blue*, or *black boxes* indicate that the comment was classified as, respectively, positive, negative, or neutral (objective). The *bottom row* presents shuffled data, i.e., the comments were arranged in a random order

of the simplest stochastic process where there is no statistical dependence between events at consecutive time-steps, and at every time-step the event probability distribution is the same. The fit diverges for large n, which means that in the data collected the probability of long clusters of the same emotional valence is large

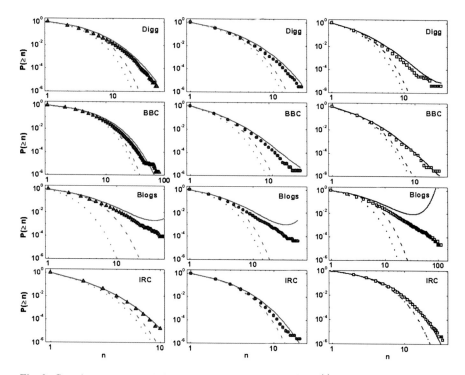

Fig. 3 Complementary cumulative distribution function (CCDF) $P^{(e)}(\geq n)$ of the cluster size for all data used in the study. *Symbols* are data (*blue triangles, red circles,* and *white squares,* respectively, for negative, positive, and neutral clusters), *dotted lines* are i.i.d. processes given by Eq. (1), *dashed lines* are Markov processes given by Eq. (2), while *solid lines* come from Eq. (3) and represent distributions based on the preferential attraction rule. The spurious increase of $P^{(e)}_{\alpha}$ $(\geq n)$ for $n \geq 40$ for Blogs data is due to violation of the scaling $p(e|ne) = p(e|e)n^{\alpha}$

Table 1 Datasets' properties

	N	T	$<e>$	Probability of emotion			Exponent α		
				$p(+)$	$p(-)$	$p(0)$	α_+	α_-	α_0
BBC	2,474,781	97,946	−0.44	0.19	0.65	0.16	0.38	0.05	0.45
Digg	1,646,153	129,998	−0.16	0.31	0.48	0.21	0.20	0.11	0.37
IRC	1,899,119	93,379	0.16	0.32	0.15	0.53	0.18	0.31	0.07
Blogs	242,057	1,219	0.14	0.35	0.22	0.43	0.23	0.19	0.16

Properties of the four datasets: number of comments N; number of discussions T; average valence in the dataset $<e>$; probability of finding positive ($p(+)$), negative ($p(-)$), and neutral emotion ($p(0)$); values of α exponents for positive (α_+), negative (α_-), and neutral (α_0) clusters

compared to the probability expected for mutually independent messages. It follows that there is a tendency for emotions of the same valence to cluster together, suggesting that there may be attractive affective forces between discussion participants—posts tend to trigger follow-up posts of the same valence.

Then, could emotions in the discussions be described by a Markov process? A Markov chain (Norris 1997) is a basic stochastic process with one memory step when the probability of the next time state depends only on the previous one by corresponding conditional probabilities. In this case,

$$P_M^{(e)}(\geq n) = p(e|e)^{n-1} \qquad (2)$$

(dotted lines in Fig. 3), where $p(e|e)$ is the conditional probability that two consecutive messages have the same emotion. It is defined as $p(e|e) = p(ee)/p(e)$, where $p(ee)$ is the joint probability of the pair ee that is measured as a number of occurrences of the two consecutive messages with the same valence e divided by the number of all appearing pairs. The fit is better than for the i.i.d., but there is still divergence for large n.

Now, let us consider conditional probability $p(e|ne)$ that after n comments with the same emotion valence the next comment will have the same valence. The data reveal the relation $p(e|e) < p(e|ee) < \ldots < p(e|ne) \approx p(e|e)n^{\alpha}$ with the characteristic exponent α representing the strength of the sublinear preferential process and α in $[0;1]$. This relationship means that finding a positive message after seven positive comments is more likely than after six. It holds true for $n < 10$, but then saturation follows, finally decreasing to zero for long clusters (see Fig. 4). Preferential processes are common in complex systems with positive loop dynamics, and they are responsible for the emergence of fat-tailed distributions, including power-law scaling (Barabási and Albert 1999; Krapivsky and Redner 2001). The influence of preferential attraction is visible in the CCDF in Fig. 3. Its decay is much slower than in the case of random or even one-step Markov processes. In order to find an analytical approximation to the cluster distribution, extending the scaling relation $p(e|ne)$ leads to the following approximation of the CCDF:

$$P_{\alpha}^{(e)}(\geq n) \approx p(e|e)^{n-1}[(n-1)!]^{\alpha} \qquad (3)$$

This approximation is presented in Fig. 3 (solid lines). The fit with the data is far better than in the case of the i.i.d. and one-step Markov distributions, especially for large n. The differences between the analytical assumption and the real data come from the artificial extension of the scaling relation $p(e|ne)$, which results in underestimation of the norm of probability and, consequently, overestimation of the number of clusters. The range of applicability of this analytical result is limited since for large n the function $P_{\alpha}^{(e)}(\geq n)$ possesses a minimum depending on the parameters $p(e|e)$ and α, and it diverges as n goes to infinity.

Figure 3 confirms that the occurrence of emotional posts cannot be described by the i.i.d. process, and there are specific correlations between emotions in consecutive posts. These correlations result from emotional interactions between discussion participants via their messages. The interactions possess an attractive character because clusters of posts with the same emotional valence are longer than clusters from random distributions. The emotion expressed by a participant depends on the emotions in previous posts—he/she tends to express emotions that have been recently used in the discussion. This observation is consistent with general ideas regarding functions of emotions (e.g., Frijda 1986). Moreover, the relations

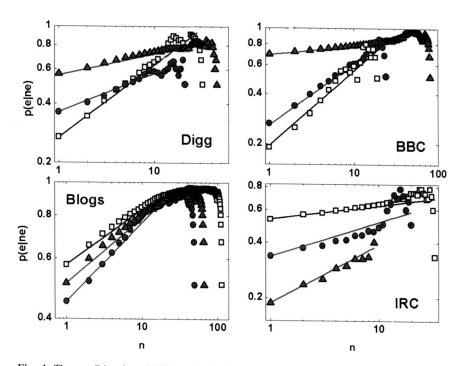

Fig. 4 The conditional probability $p(e|ne)$ of the next comment occurring having the same emotion for Digg, BBC, Blogs, and IRC data. *Symbols* are data (*blue triangles*, *red circles*, and *white squares*, respectively, for negative, positive, and neutral clusters) and *lines* reflect the fit to the preferential attraction relation $p(e|ne) = p(e|e)n^{\alpha}$

observed in both Figs. 3 and 4 indicate that the behavior of the participants can be regarded as a collective one—the more the emotional posts submitted the more emotional will be the next one.

Note that the collective effect happens regardless of emotion valence—in other words, the affective attractive forces that produce a snowball effect in a set of consecutive messages are not sensitive to the emotion type. The prerequisite for a significant value of such a force is a large *number of messages* in the preceding cluster of homogeneous emotions. This observation supports recent findings by Sabucedo et al. (2011) who showed that both positive and negative emotions (respectively, enthusiasm and anger in their study) can be responsible for triggering collective actions (political demonstrations).

4   Life-Span of the BBC Forum and Digg Communities

The influence of emotions on the duration of BBC forum and Digg discussions was investigated as follows. Threads of the same size were grouped together, and a moving average of the emotion type of the last ten comments was calculated for

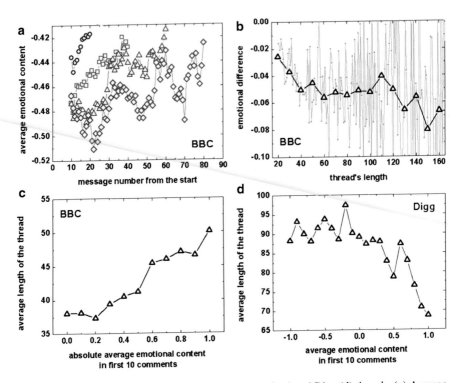

Fig. 5 Time dependence of emotions in BBC Forum (**a**, **b**, **c**) and Digg (**d**) threads. (**a**) Average emotion valence in the thread (moving average of the previous 10 messages in the thread). Four groups of threads of lengths 20, 40, 60, and 80 are represented by different symbols (respectively *circles*, *squares*, *triangles*, and *diamonds*). Shorter threads start from emotional levels closer to zero. (**b**) Emotional level (valence) at the beginning of a thread minus the emotional level at the end as a function of thread length (*grey symbols*). *Black triangles* display binned data. Longer threads use more emotional "fuel" over time. (**c**) Average length of the thread as a function of the absolute value of the average emotion valence of the first 10 comments. Emotional thread starts, whether positive or negative, usually lead to longer discussions. (**d**) Average length of thread in Digg data as a function of the average valence of the first 10 comments. Here, only negative start of thread leads to longer discussions

each point. As seen in Fig. 5a, shorter threads tend to start from a lower (i.e., less negative) emotional level than longer ones. On the other hand, threads end with a similar mean emotional valence value regardless of their lengths—the last point of each data series in 3a (circles, squares, triangles, and diamonds) is at almost the same level, i.e., about −0.42. This phenomenon is echoed in Fig. 5b where the average emotional valence of the first ten comments minus the average emotional valence of the last ten comments is plotted, showing that longer threads have bigger eventual decreases in negative valence. Figure 5c also suggests that the initial emotional content (whether positive or negative) may be used as an indicator of the expected length of a thread—low absolute average emotion valences lead to shorter discussions. A possible heuristic explanation is that the first few posts in a thread may give it the potential (emotional fuel) to propel further discussion. Once the emotions driving the discussion dry out, the thread is no longer of interest to its participants and it may die. For the threads possessing higher initial levels of emotion, it takes more comments to resolve the emotional issue, resulting in longer threads. A similar although slightly different phenomenon is spotted in the Digg data. Here, as seen in Fig. 5d, only negative start of the thread prolongs the discussion.

5   Users Impact on the Discussion in the BBC Forum

Here we consider user activity a_i defined to be the total number of posts written by user i in all discussion threads during the observation period. For simplicity, this quantity will also be referred to as a. The maximum observed activity in the dataset is $a_{max} = 18274$, i.e., one user authored more than 18,000 messages, while the average activity is $<a> = 137$ and the median $m_a = 3$. The number of occurrences of a is presented in Fig. 6a (red triangles) and it is well fitted by the power-law relationship $h_a \sim a^{-\beta}$, $\beta = 1.4$ (black line in Fig. 6a). The relatively small value of the exponent β suggests a high number of very active users of this forum. Since all discussions in the forum are split into separate threads j, we define d_{ij} (or d for short) to be the local activity of user i in thread j measured by the number of posts that this user submitted in the discussion. Whereas both its maximum and average values (respectively $d_{max} = 1582$ and $<d> = 2.84$) are lower than in the case of a, the number of occurrences of d shown as black circles in Fig. 6a still follows a similar relationship $h_d \sim d^{-\gamma}$ with exponent $\gamma = 2.9$ (red line in Fig. 6a), which is double that of a.

While user behavior shows a strong tendency to be scale-invariant, this is not so clear for the thread statistics shown in Fig. 6b. Here, we consider thread length L and the number of unique users U posting at least one comment in the thread. Histograms of both quantities h_L (black circles) and h_U (red triangles) display power-law tails for $U,L > 20$. This is most prominent in the case of h_U, which is also characterized by a rather large exponent $\eta = 4.9$ (black line in Fig. 6a).

To understand the impact exerted by the most frequent users on the length of a thread, consider the dependence between the normalized number of unique users in a single thread defined by $u = U/L$ and thread length (Fig. 6c). For short threads (L between 1 and 10), u is about 0.6–1 while for threads larger than 400 comments, it drops below 0.1. A good fit is $u(L) = A(L+b)^{-0.58}$ (the blue line in Fig. 6c); thus, the number of unique users grows more slowly than linearly with thread lengths. This suggests that mutual discussions between specific users rather than a large number of independent comments submitted by many users sustain thread life.

The following quantities describe the emotions of individual debaters and discussions threads. The average (global) emotion of a user e_a is the sum of all emotions e in posts written by the user i divided by his/her activity a_i. The average emotion of a thread e_L is the sum of all emotions in the thread j divided by its length

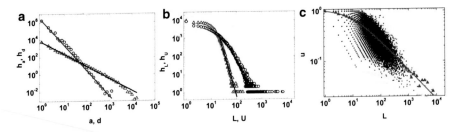

Fig. 6 (a) Histogram of user activity a (*triangles*); histogram of user activity in thread d (*circles*). Lines are fits to the data and they follow relations $h_a \sim a^{-\beta}$ and $h_d \sim d^{-\gamma}$ with $\beta = 1.4$ and $\gamma = 2.9$. (b) Histogram of threads with length L (*circles*); histogram of the number of unique users U making a comment in the thread (*triangles*). The *black line* is a fit to the tail of the distribution and it follows the relation $h_U \sim U^{-\eta}$ with $\eta = 4.9$. (c) The normalized number of unique users u making a comment in a thread of length L (*dots*—original data, *triangles*—data binned logarithmically). The *blue line* corresponds to relation $u = A(L + b)^{-\delta}$ with fitted parameters $\delta = 0.58$, $A = 3.72$ and $b = 8.6$

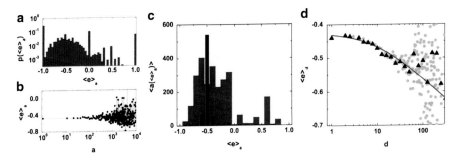

Fig. 7 (a) Probability distribution of users' global average emotions $\langle e \rangle_a$. (b) Users' global average emotion $\langle e \rangle_a$ versus users' global activity a. (c) Users' average global activity $\langle a (\langle e \rangle_a) \rangle$ versus their global average emotion $\langle e \rangle_a$: red bars—empirical data, black bars—shuffled data. (d) Relationship between users' average emotion in a thread $\langle e \rangle_d$ and users' activity in the thread d. *Grey circles* are original data, *black triangles* are binned data, and the *red curve* corresponds to equation $\langle e \rangle_d = A_1 + B_1 \ln(d + b)$ with $A_1 = -0.31$, $B_1 = -0.054$ and $b = 8.6$

L_j. The third value e_d is the average emotional expression of the user i in the thread j. The main features of the distribution $p(e_a)$, presented in Fig. 7a, are peaks for $e_a = -1, 0, 1$ which are a straightforward effect of the large number of users with $a = 1$ and threads with $L = 1$ (see Fig. 6a, b). The local maximum around $e_a = -0.5$ is a specific attribute of the BBC Forum because it possesses a strong bias toward negative emotions, with an average value of $\langle e \rangle = -0.44$. We observe similar distribution shapes for $p(e_L)$ and $p(e_d)$.

So far we have treated user activities and emotions as mutually independent variables, but we now consider the relationship between them. Figure 7b plots users' global average emotions $\langle e \rangle_a$ versus global activity a. Neglecting fluctuations for large values of a caused by small numbers of very active users, there is a

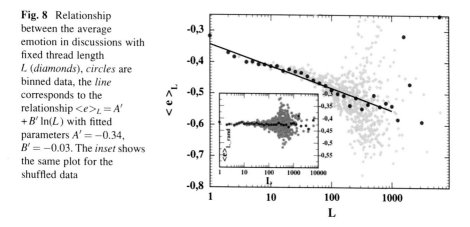

Fig. 8 Relationship between the average emotion in discussions with fixed thread length *L* (*diamonds*), *circles* are binned data, the *line* corresponds to the relationship $<e>_L = A' + B' \ln(L)$ with fitted parameters $A' = -0.34$, $B' = -0.03$. The *inset* shows the same plot for the shuffled data

constant mean emotion that is around the forum's average value $<e>$. Hence, on average, the user activity level a does not influence his/her average emotions $<e>$. In Fig. 7c, the reversed relationship is plotted, i.e., the average global activity versus users' average emotions (red bars). For comparison, we present shuffled data (black bars) where the emotional values of posts were randomly interchanged between users. Whereas the second distribution follows a Gaussian-like function, the original set is characterized by a broad maximum stretching across almost all of the negative part of the plot and some minor fluctuations in the positive part. This means that although there are the same mean emotions for groups of users of various activities (see Fig. 7b), there are different average activities for users of various mean emotions.

Users can take part in many threads; thus, their local and global activities as well as corresponding local emotions can be very different. But how are users' emotions $<e>_d$ expressed in a thread connected to the activity level d in it? Figure 7d shows the average emotions of a user in a thread as a function of the user's local activity. In this case, an increase in activity in a particular thread leads to more negative average emotions in the thread. Recall that there was no relationship between a user's global activity and his/her emotions, as shown in Fig. 7b. For longer discussions, there is a more homogeneous group of users (see Fig. 6c); thus, on average one user writes a larger number of posts $<d>(L) = 1/u(L)$. As shown in Fig. 7d, the average emotions for users locally more active decreases logarithmically. These two effects cause the longer threads to possess, on average, more negative emotions. In fact in Fig. 8, there is a logarithmic decay in mean thread emotions $<e>_L$ as a function of thread length L. To confirm the statistical validity of this phenomenon, we randomly shuffled emotions between various threads. The inset in Fig. 8 proves that in this case mean thread emotion is independent of thread length. The qualitative outcome of Fig. 8 resembles Fig. 5b with respect to the idea of emotional (negative) fuel that has to be included in the discussion in order to sustain it.

6　Conclusions

On the basis of automatic sentiment detection methods applied to huge datasets, we have shown that Internet users' messages correlate at the simplest emotional level—positive, negative, or neutral messages tend to provoke similar responses. The collective character of the emotions expressed was evident in several different types of e-community—it was observed for BBC forums and Digg (mainly negative emotions), for the Blogs (mainly positive comments), and also in IRC dialogues (neutral). The strength of emotional interactions can be indirectly measured by the parameter α expressing the influence of the most recent emotional cluster on the probability that the next post has the same emotion. The results indicate the presence of online collective behavior among users that creates longer discussion threads.

We also found patterns in individual users' emotional behaviors in online BBC Forums. We observed a scale-free distribution of users' activities in the whole forum and in singular threads as well as power law tails for the distribution of thread lengths and the number of unique users in a thread. At the level of the entire forum, negative emotions boost users' activities; participants with more negative emotions write more posts. At the level of individual threads, users that are more active in a specific thread tend to express more negative emotions in it and seem to be the key agents for sustaining discussion. As result, longer threads possess more negative emotional content. Overall, then, negativity is the key sentiment to start and sustain online discussions, at least in the forums investigated here.

Acknowledgments　This work was supported by a European Union grant by the 7th Framework Programme, Theme 3: Science of complex systems for socially intelligent ICT. It is part of the CyberEmotions (*Collective Emotions in Cyberspace*) project (contract 231323). J.A.H, A.Ch. and J.S. acknowledge support from Polish Ministry of Science Grant 1029/7.PR UE/2009/7.

References

Barabási A-L (2005) The origin of bursts and heavy tails in human dynamics. Nature 207:435–433

Barabási A-L, Albert R (1999) Emergence of scaling in random networks. Science 286:509–519

Borge-Holthoefer J, Rivero A, Moreno Y (2012) Locating privileged spreaders on a online social network. Phys Rev E 85:066123

Castellano C, Fortunato S, Loreto V (2009) Statistical physics of social dynamics. Rev Mod Phys 81:591–646

Chmiel A, Kowalska K, Hołyst JA (2009) Scaling of human behavior during portal browsing. Phys Rev E 80:066122

Chmiel A, Sobkowicz P, Sienkiewicz J, Paltoglou G, Buckley K, Thelwall M, Holyst JA (2011a) Negative emotions boost user activity at BBC forum. Phys A 390:2936–2944

Chmiel A, Sienkiewicz J, Thelwall M, Paltoglou G, Buckley K, Kappas A, Hołyst JA (2011b) Collective emotions online and their influence on community life. PLoS ONE 6(7):e22207

Dodds PS, Danforth CM (2010) Measuring the happiness of large-scale written expression: songs, blogs, and presidents. J Happiness Stud 11:441–456

Feller W (1968) An introduction to probability: theory and its applications. Wiley, Hoboken, NJ

Frijda NH (1986) The emotions. Cambridge University Press, Cambridge, MA

Gamon M, Aue A, Corston-Oliver S, Ringger E (2005) Lect Notes Comput Sci 3646:121–132

Huberman BA, Pirolli PLT, Pitkow JE, Lukose RJ (1998) Strong regularities in World Wide Web surfing. Science 280:95–97

Krapivsky P, Redner S (2001) Organization of growing random networks. Phys Rev E 63:066123

Kujawski B, Hołyst JA, Rodgers GJ (2007) Growing trees in internet news groups and forums. Phys Rev E 76:036103

Lang PJ, Davis M (2006) Emotion, motivation, and the brain: reflex foundations in animal and human research. Prog Brain Res 156:3–34

Macdonald C, Ounis I (2006) The TREC Blogs06 collection: creating and analyzing a blog test collection (Technical Report TR-2006-224). Department of Computer Science, University of Glasgow, Glasgow

Mitrović M, Paltoglou G, Tadić B (2010) Networks and emotion-driven user communities at popular blogs. Eur Phys J B 77:597–609

Norris JR (1997) Markov chains. Cambridge University Press, Cambridge, MA

Onnela J-P, Reed-Tsochas F (2010) Spontaneous emergence of social influence in online systems. Proc Natl Acad Sci USA 107:18375–18380

Paltoglou G, Thelwall M, Buckely K (2010) Online textual communication annotated with grades of emotion strength. In: Proceedings of the third international workshop on EMOTION (satellite of LREC): corpora for research on emotion and affect, Valletta, Malta, pp 25–31

Pang B, Lee L (2008) Opinion mining and sentiment analysis. Found Trends Inf Retr 1(1–2):1–135

Pohorecki P, Sienkiewicz J, Mitrović M, Paltoglou G, Hołyst JA (2013) Statistical analysis of emotions and opinions at Digg website. Acta Phys Pol 123:604–614

Postmes T, Brunsting S (2002) Collective action in the age of the internet mass communication and online mobilization. Soc Sci Comput Rev 20:290–301

Radicchi F (2009) Human activity in the web. Phys Rev E 80:026118

Riloff E, Patwardhan S, Wiebe J (2006) Feature subsumption for opinion analysis. In: Proceedings of the conference on empirical methods in natural language processing, Morristown, NJ, USA, pp 440–448

Sabucedo JM, Durán M, Alzate M, Barreto I (2011) Emotions, ideology and collective political action. Univ Psychol 10:27–34

Sebastiani F (2002) Machine learning in automated text categorization. ACM Comput Surv 34(1): 1–47

Sienkiewicz J, Skowron M, Paltoglou G, Hołyst JA (2013) Entropy-growth-based model of emotionally charged dialogues. Adv Complex Syst 16:1350026

Sobkowicz P, Sobkowicz A (2010) Dynamics of hate based Internet user networks. Eur Phys J B 73:633–643

Stürner S, Simon B (2004) Collective action: towards a dual pathway model. Eur Rev Soc Psychol 15:59–99

Szell M, Lambiotte R, Turner S (2010) Multirelational organization of large-scale social networks in an online world. Proc Natl Acad Sci USA 107:13636–13641

Taylor V (1995) Watching for vibes: bringing emotions into the study of feminist organizations. In: Ferree MM, Martin PY (eds) Feminist organizations: harvest of the new women's movement. Temple University Press, Philadelphia, pp 223–233

Tumasjan A, Sprenger TO, Sandner PG, Welpe IM (2010) In: Proceedings of the fourth international AAAI conference on weblogs and social media. AAAI Press, Menlo Park, CA, pp 178–185

van Zomeren M, Postmes T, Spears R (2008) Toward an integrative social identity model of collective action: a quantitative research synthesis of three socio-psychological perspectives. Psychol Bull 134:504–535

van Zomeren M, Postmes T, Spears R (2012) On conviction's collective consequences: integrating moral conviction with the social identity model of collective action. Br J Soc Psychol 51:52–71

Walther J, Parks M (2002) In: Knapp M, Daly J, Miller G (eds) The handbook of interpersonal communication. Sage, Thousand Oaks, CA, pp 529–563

Weroński P, Sienkiewicz J, Paltoglou G, Buckley K, Thelwall M, Hołyst JA (2012) Emotional analysis of blogs and forums data. Acta Phys Pol A 121:B128–B132

Wilson T, Wiebe J, Hoffman P (2009) Recognizing contextual polarity: an exploration of features for phrase-level sentiment analysis. Comput Linguist 35(3):399–433

Wright SC, Taylor DM, Moghaddam FM (1990) Responding to membership in a disadvantaged group: from acceptance to collective protest. J Person Soc Psychol 58:994–1003

Evaluation of Media-Based Social Interactions: Linking Collective Actions to Media Types, Applications, and Devices in Social Networks

Alan Keller Gomes and Maria da Graça Campos Pimentel

Abstract There is a growing number of opportunities for users to perform collective actions in social networks: Such collective actions engage users in correspondents social interactions. Although some models for representing users and their relationships in social networks have been proposed, to the best of our knowledge, these models do not explain what the underlying social interactions are. In previous work, we have proposed a human-readable technique for modeling and measuring social interactions, which resulted from users' actions that involved, for instance, media types, interaction devices, and viral content. In our technique, social interactions are represented as behavioral contingencies in the form of *if-then* rules, which are then measured using an established data mining procedure. After being able to represent and measure a variety of social interactions, we identified the opportunity of transforming our technique into a method for capturing, representing, and measuring collective actions in social networks. In this chapter, we present our method and detail how it was applied to represent and measure social interactions among a group of 1,600 Facebook users over the period of 7 months. Our results report the link among actions (e.g., *like*), media objects (e.g., photo), application type (Web or mobile), and device type (e.g., Android).

A.K. Gomes (✉)
Federal Institute of Goiás—IFG, College of Inhumas, Inhumas, GO, Brazil

University of São Paulo, Institute of Mathematics and Computer Science, São Carlos, SP, Brazil
e-mail: alankeller@ifg.edu.br; alankeller@icmc.usp.br

M. da Graça Campos Pimentel
University of São Paulo, Institute of Mathematics and Computer Science, São Carlos, SP, Brazil
e-mail: mgp@icmc.usp.br

N. Agarwal et al. (eds.), *Online Collective Action*, Lecture Notes in Social Networks, 75
DOI 10.1007/978-3-7091-1340-0_5, © Springer-Verlag Wien 2014

1   Introduction

Social networks allow the engagement, communication, sharing, realization of collaborative activities, and social interactions among users. Many alternatives for accessing social networks through Web, tablet, and smartphone applications allow users to experience new types of social interaction (Bentley and Metcalf 2009) (Cowan et al. 2011).

Social interactions have been defined as the acts, actions, or practices of two or more people mutually oriented toward each other (Rummel 1976). Results from behavioral sciences argue that social interactions may be specified as *behavioral contingencies* in the form of *if-then* rules, which correspond to observations of what people do, or do not do, in a variety of situations (Mechner 1959). Results from sociology suggest that the *relationships* among individuals may be modeled as graphs (Granovetter 1973; Holland and Leinhardt 1970), and in social networks, the user *interaction* has been represented as a directed graph called *interaction graph* (Wilson et al. 2009).

A social network may be defined as a set of social entities (actor, points, nodes, or agents) that may have relationships (edges or ties) with one another. In the social network analysis research field, social networks are modeled as graphs (or sociograms) (Freeman 2004; Scott 2000; Wasserman and Faust 1994). As in social environments, users are social entities, and their relationships may be bidirected (i.e., friend) or directed (i.e., following).

In social network analysis, models for the evaluation of user interaction are usually based on graph theory (Mislove et al. 2007; Tan et al. 2011) and, according to the small-world principle (Kleinberg 2000; Watts 1999), are investigated using data mining techniques such as clustering (Abrol and Khan 2010; Negoescu et al. 2009), classification (Bonchi et al. 2011; Ekenel and Semela 2011), and prediction (Jin et al. 2010; Shi et al. 2011).

Considering the importance of understanding the underlying social interactions in social media environments, in previous work we have proposed a human-readable *if-then* rule–based technique for the representation and evaluation of social interactions in social networks (Gomes and Pimentel 2011d). Our technique combines the representation of social interactions as *if-then* rules in a behavioral contingency language with the evaluation of behavioral contingencies by means of data mining procedures.

As an example, upon observing a particular social interaction involving users a and b that perform actions A_1 and A_2 leading to consequence C_1, this interaction may be registered as the rule $aA_1 \cap bA_2 \rightarrow abC_1$. A set of such rules, extracted from observing a particular social interaction, is used in qualitative evaluations relative to the social interaction itself. For instance, in game-setting behavioral contingencies, *if-then* rules may be analyzed to determine how the game is played (Mechner 2008).

We have applied our technique to study many social interactions among Facebook users (Gomes and Pimentel 2011a, c, 2012a, b) as well as to study collaborative scenarios (Gomes et al. 2011; Gomes and Pimentel 2011b) supported by a social approach for the *Watch-and-comment* paradigm (Fagá et al. 2010). This chapter contributes to the proposed research topic by presenting a method for capturing, representing, and measuring collective actions in social networks. In regard to validating the method, we detail how it was applied to represent and measure the social interaction among a group of 1,600 Facebook users over a period of 7 months. Our results report the use of actions (e.g., *like*), media objects (e.g., *photo* or *link*), application type (e.g., Web or mobile), and device type (e.g., Android).

The chapter is organized as follows: Sect. 2 discusses related work; Sect. 3 revisits our human-readable technique for representing and measuring social interactions; Sect. 4 details the new proposed method; Sect. 5 reports the results of a study involving a group of Facebook users; and Sect. 6 presents our final remarks.

2 Related Work

Related work investigates the computation of behavioral sciences, the evaluation of sociability in social media systems, and the modeling of interaction among users in Twitter and Facebook.

The computation of behavioral sciences is an emergent research field, which permits processing several nonverbal behavioral cues including facial expressions, body postures and gestures, and vocal outbursts such as laughter. Boulard et al. (2009) propose a set of recommendations for enabling the development of the next generation of socially aware computation. A range of simple techniques to access personal information relevant to social research matters on the Web are presented by Wilkinson and Thelwall (2011). However, the researchers do not build a model for representation and evaluation of social interactions among users.

Sociability-related issues play an important role in the design of social media applications in the scenario of mobile digital TV, as demonstrated by Geerts (2010) and Chorianopoulos (2010). In the context of evaluation of social TV applications, Geerts and Grooff (2009) heuristics and guidelines—established to support the assessment of social skills in computer systems—aimed at social TV and social video. Such evaluations do not consider the underlying social interactions when computing quantitative measures.

Based on a combination of a user's position, polarity of opinion, and textual quality of *tweets*, Bigonha et al. (2010) propose a technique for ranking users as evangelists and detractors. They made a topological analysis of the network by using typical measures and by analyzing *retweets* and *replies*. By using Twitter as a test bed, Choudhury et al. (2011) propose an iterative clustering technique for selecting a set of items on a given topic that matches a specified level of diversity. They also observed that content was perceived to be more relevant when it was either highly homogeneous or highly heterogeneous.

Both the above-mentioned research papers make use of textual analysis of the messages shared among Twitter users. In contrast with the work presented in this paper, research carried out here does not represent users' activities (*tweets*, *retweets*, and *replies*) as actions, and it does not consider any type of media that can be shared in the messages apart from text. Furthermore, the models employed in the presentation and in the analysis of results do not focus on a human-readable interpretation.

The interaction among Facebook users has been studied by Wilson et al. (2009), and results have shown that a social network user interaction graph is a subset of a social graph. Using data from Facebook, Backstrom et al. (2011) found that the balance of attention is a relatively stable property of an individual over time and that there is an interesting variation across both different groups of people and different modes of interaction.

These research papers have used Facebook data and analyzed users' activities, but they do not analyze user behavior, action, and media. Just as the research that uses Twitter data, the models employed in the presentation and in the analysis of results do not explicit how user engagement in social interactions takes place.

In the next section, we revisit our proposed human-readable technique for representing and evaluating media-based social interactions.

3   A Human-Readable Technique for Representing and Evaluating Social Interactions

Our technique for representing and evaluating social interactions uses the Mechner language for representing social interactions as *if-then* rules and data mining procedures to evaluate the rules.

3.1   Social Interactions in the Mechner Language

In behavioral science, any kind of social interactions can be specified as a conditional relationship in the form of *if-then* statements, e.g., *behavioral contingencies* (Skinner 1953). For example,

- A given law may be written as a rule such as "If a person does or does not perform a certain act, certain consequences for that person will follow." In essence, laws are behavioral contingencies intended to regulate, modify, or influence behavior.
- Game rules, e.g., tic-tac-toe, are behavioral contingencies that determine how the game is played.

Mechner (1959) presented one of the first notation systems for codifying any behavioral contingency which combined Boolean algebra with a set of diagrams. Weingarten and Mechner (1966) detailed Mechner's original work by representing social interactions as independent variables in the form of *if-then* rules. More recently, Mechner (2008) presented a formal symbolic language for codifying any behavioral contingencies which involved several participants. In the *Mechner language*, behavioral contingencies are logic implications that can be evaluated as independent variables. Some important elements of this language are

1. *Action (or actions)*: matching the antecedent of the contingency, i.e., $A \rightarrow$. If there is more than one action, they are represented as $A_1 \cap A_2 \cdots \rightarrow$.
2. *Agent(s) of action(s)*: represented by lowercase letters placed in front of an A. For example, agent a performed action A, i.e., aA. One letter can represent a single agent or a group of agents that perform an action.
3. *Consequence*: corresponds to the consequence of the contingency, i.e., $\rightarrow C$. If there is more than one consequence, they are represented as $\cdots \rightarrow C_1 \cap C_2$.

For example, behavioral contingencies codified in the form of an *if-then* statement using the Mechner language is:

- $\bar{a}A_1 \cap bA_2 \rightarrow \bar{a}bC_2$. *If a does not execute action A_1 and b executes action A_2, then* the consequence C is not perceived by a but is perceived by b.
- $aA_1 \cap bA_2 \rightarrow aC_1 \cap bC_2$. *If a executes action A_1 and b executes action A_2, then* the consequence C_1 is perceived by a, and the consequence C_2 is perceived by b.

The action that starts the social interaction, which is the case of A_1 in the above example, is called *social stimulus* (Skinner 1953). Although other notation systems have been proposed to codify behaviors in experimental analysis processes (e.g., Mattaini 1995), in our work we use the Mechner language (Mechner 2008) for representing behavioral contingencies as Boolean expressions in disjunctive normal form, i.e., implications within $-$, $\cap$(*not, and*) connectives. This mathematical property is necessary for the data mining procedure we have adopted, which we describe next.

3.2 Behavioral Contingencies Representation

In order to use the Mechner language to represent situations involving social network users in social interactions, we have to identify the Mechner language elements, i.e., *actions A, agents of actions* (e.g., user a, or group (of users) k and l), and *consequences C*.

Each social network has a specific set of *actions A, agents of actions*, and *consequences C*. Figure 1 presents some actions that can be performed in social networks. As examples of *social stimuli*, we have on Facebook: $A_1 = $ post on one's Wall; on Twitter: $A_1 = $ Tweet (post a text message with maximum 140 characters); on Google+: $A_1 = $ post news.

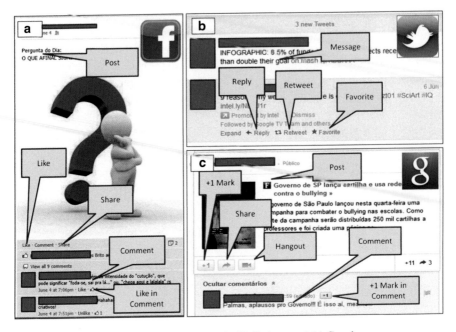

Fig. 1 Actions in social networks: (**a**) Facebook; (**b**) Twitter; and (**c**) Google+

Users in a social network are *agents of actions*, and they can perform one or more actions individually (e.g., user a or b) or in groups (e.g., group k or l) according to permissions provided by the social network. As a result, users may be notified of one or more *consequences C* of other users' actions. Moreover, depending on the permission they have, users may also act as results of other users' actions. For example, user b can *like a post* (C_1) or can *comment a post* (C_2) after being notified that user a *posted on his wall*.

After identifying *actions*, *agents of actions* (users), and *consequences*, we represent social interactions. For example,

- *if* a Facebook user a performs the action $A_1 = post$ *a message* on his Wall,

  - *then* user a and user b $C_1 = are$ *notified of this post*,
  - *then if* user b the action $A_2 = like$ *that post* (after being notified of the posting),
  - *then* user a and user b $C_2 = are$ *notified of this like*,

Using the Mechner language, we represent this social interaction as $aA_1 \rightarrow abC_1 \rightarrow bA_2 \rightarrow abC_2$, e.g., $aA_1 \cap bA_2 \rightarrow abC_1 \cap abC_2$.

When modeling behavioral contingencies, the identification of a user or a group of users and the granularity of actions and consequences are defined by the analyst with support from specialists to identify social interactions.

3.3 Behavioral Contingency Measurement

In our work, behavioral contingencies codified in the Mechner language as *if-then* statements are represented in the form $Body \rightarrow Head$ (in short, $B \rightarrow H$). For example, considering $B = aA_1 \cap bA_2$ and $H = abC_1 \cap abC_2$, an *if-then* behavioral contingency is represented as $R = aA_1 \cap bA_2 \rightarrow abC_1 \cap abC_2$.

In the Rule Learning (Fürnkranz et al. 2011) research field, *if-then* rules are general implications in the form of $B \rightarrow H$, which can be evaluated by a variety of measures (Azevedo and Jorge 2007; Lavrac et al. 1999) such as *confidence*, *support*, and *cosine correlation* (Han and Kamber 2005; Martínez-Ballesteros and Riquelme 2011). These measures can be used in quantitative evaluations.

Using a data mining procedure, the value of a rule $B \rightarrow H$ can be measured by comparing it with a set of observations (Lavrac et al. 1999). For example, the number n of behavioral contingencies observed during a particular social experience can be computed using classic data mining contingency values $bh, b\overline{h}, \overline{b}h, \overline{b}\overline{h}$ as follows:

$$n = bh + b\overline{h} + \overline{b}h + \overline{b}\overline{h}$$

where

bh is the number of observed situations for which head b and body h are true

$b\overline{h}$ is the number of observed situations for which the body b is true and the head h is false

$\overline{b}h$ is the number of observed situations for which the body b is false and the head h is true

$\overline{b}\overline{h}$ is the number of observed situations for which head b and body h are false

As an application of the mapping of Mechner contingencies into data mining rules, contingency values can be used to calculate measures of a given rule in a set of observations. Table 1 details how the contingency values bh, $b\overline{h}$, $\overline{b}h$, and $\overline{b}\overline{h}$ are used in the computation of four classic rule evaluation measures (Martínez-Ballesteros and Riquelme 2011):

- The *Support* measure for a rule R determines the applicability of such a rule to a given set of observations, which in turn determines how frequently H and B will appear in the set of observations. This measure reflects the usefulness of the discovered rules.
- The *Confidence* measure for a rule R computes the reliability of the inference made by rule R, thus determining how frequently H appears in observations that contain B. This measure reflects the certainty of the discovered rules.
- The *Cosine correlation* measure for a rule R determines the strength (or weakness) of the association between B and H.
- The *Leverage* measure for a rule R computes the proportion of additional cases covered by both B and H and those cases in which expected B and H were independent of one another.

Table 1 Rule evaluation measures

Support	Confidence	Cosine correlation	Leverage
$\dfrac{bh}{n}$	$\dfrac{bh}{b}$	$\dfrac{bh}{n * \sqrt{\frac{b*h}{n^2}}}$	$\dfrac{bh}{n} - \left(\dfrac{h}{n} * \dfrac{b}{n}\right)$

Measures for rule evaluation can be *symmetric* or *asymmetric* (Tan et al. 2005). The measures Support (SupR), Cosine Correlation (CosR), and Leverage (LevR) are *symmetric* measures because their values are identical for rules $B \rightarrow H$ and $H \rightarrow B$. In contrast, Confidence (ConR) is asymmetric because its values may not be the same for rules $B \rightarrow H$ and $H \rightarrow B$. Symmetric measures are generally used for evaluating B and H as independent values, while asymmetric measures are more suitable for analyzing rules, i.e., rules involving B and H. Conventionally, these measures are represented as 0–100 % values rather than 0–1 (Han and Kamber 2005).

3.4 Experiences from Applying our Technique

We first studied contingencies as social interactions associated with the asynchronous sharing of video links and annotation sessions (Gomes et al. 2011) and the synchronous and asynchronous sharing of collaborative annotations (Gomes and Pimentel 2011b) on YouTube videos.

We then applied the approach to analyze social interactions on Facebook in order to identify the social situations in which users are most involved (Gomes and Pimentel 2011c). In the context of our research, we identified the need for a tool which allowed both the description and the evaluation of behavioral contingencies, which then led us to propose a human-readable *if-then* rule–based technique for the representation and evaluation of social interactions in social networks (Gomes and Pimentel 2011d). As result, the technique guides a researcher on how to combine the Mechner language and rule-based data mining procedures in order to carry out the description and the evaluation of social interactions.

The technique was applied to study social interactions where users provide media objects via smartphones (Gomes and Pimentel 2011a) as well as interactions in which users make use of media servers to provide media objects (including YouTube and Soundcloud) (Gomes and Pimentel 2011e).

Building upon a recurring sequence of the steps employed in the application of the technique, we detailed an interactive and iterative procedure to apply such a technique and used it to identify viral content shared on Facebook (Gomes and Pimentel 2012b). Through such the procedure, it was also possible to identify, among the everyday social interactions in a particular country, those where Facebook users were engaged in social manifestations against corruption (Gomes and Pimentel 2012a).

After applying the procedure, we observed the need to further specify the executed steps, which will be discussed next.

4 A Method for Representing and Measuring Social Interactions

As summarized in the previous section, the technique we have been developing has been successful in allowing the identification of a variety of details involved in interactions among users of social networks. In order to provide a more detailed guidance on how the procedure can be replicated, in this section we propose a method which structures both the steps involved in the analysis and the inputs and outputs in each step.

This method aims at detailing actions, media objects, and application and device types within rules. The method comprises six phases: Capturing, Representation, Measurement, Interpretation, and Specialization. From one phase to the other, output generated is used as input for the following phase: a list of attributes of interest, raw data set, projected data sets, sets of rules, rule measures, and new specified attributes. Given the exploratory character of the investigation, the method is iterative. Next, we will detail each of the phases. Figure 2 presents an overview of the method.

4.1 Capturing

Data capturing happens automatically, for example, when using APIs (Gomes and Pimentel 2011a, c, 2012b) or when analyzing logs generated from the use of social media systems (Gomes et al. 2011; Gomes and Pimentel 2011b). The identification, selection, and preparation of the data to be captured are typical activities of this stage and take into account observations on the social interaction. Users' personal information must be collected (name, address, sex, preferences, etc.). Other necessary information includes the content of a post, the server which provides the media object within a post, the international standard language, the URL, the source, the caption, the description, the timestamps, the location available for each post, etc.

- *Input*: social media environment; a list of attributes of interest: actions, consequences, media types, user identification, language, URL, caption, description, timestamps, location, etc., captured via API or log analysis
- *Output*: raw data, description of social interactions

4.2 Preparation

All collected data can be cleaned, selected, and/or transformed in order to be used in the subsequent phases of representation and measurement of social interactions. In this phase, the identification of the elements of the Mechner language (actions,

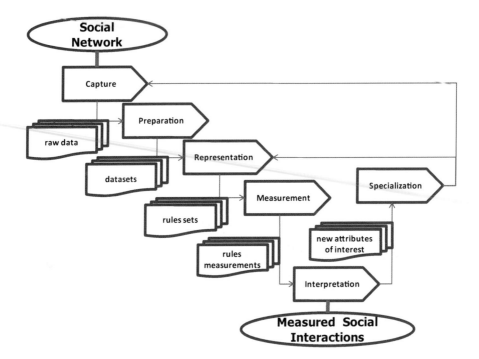

Fig. 2 An overview of our six-phase method: from social networks to measured social interactions

users, and consequences) is needed (Gomes and Pimentel 2011b, 2012b). For example, specified attributes can be the media type posted in a message, users' actions performed around the media object posted, or the notification of each posting or action performed.

- *Input*: raw data, description of social interactions
- *Output*: projected data set with actions, action agent, and consequences

4.3 Representation

Social interactions are represented as behavioral contingencies, in the form of *if-then* rules, in scenarios which involve Facebook users in social interactions (Gomes and Pimentel 2011c, d) provided via smartphone applications (Gomes and Pimentel 2011a), and in social interactions which spread viral content (Gomes and Pimentel 2012b) and social manifestations against corruption in Brazil (Gomes and Pimentel 2012a). The manual acquisition of behavioral contingencies can be made from observing social situations in the social media system, and their representation is made with support from specialists in social interactions after identifying the elements of the Mechner language.

- *Input*: projected data set with actions, action agents, and consequences
- *Output*: sets of rules

4.4 Measurement

Each social interaction described as an *if-then* rule is represented in the general form $B \rightarrow H$ for computing bh, $b\overline{h}$, $\overline{b}h$, and $\overline{b}\overline{h}$ values. After that, each rule is measured by using computing rule evaluation measures. The measures *Support*, *Confidence*, and *Cosine Correlation* are used to measure media-based social interactions (Gomes and Pimentel 2011a, c, d, 2012a) and *Sensitivity* and *Laplace* to measure social interactions which spread viral content (Gomes and Pimentel 2012b) on Facebook.

- *Input*: sets of rules
- *Output*: rule measurements

4.5 Interpretation

The measurements must be interpreted by analyst or specialist users who are interested in analyzing social interactions. If the results are sufficient, the process is completed at this stage. Otherwise, the rule specialization can be carried out, and the process can be repeated from the phases Representation or Preparation. For example, underlying social interactions can be specialized in media-based social interactions by (1) detailing media types within social interactions (Gomes and Pimentel 2011c, d), (2) detailing smartphone user applications (Gomes and Pimentel 2011a), (3) server content providers (Gomes and Pimentel 2012b) and specific bags of words to feature social movements (Gomes and Pimentel 2012a) within media-based social interactions.

- *Input*: rule measures
- *Output*: social interaction measurements (completed process) or a new list of attributes of interest

4.6 Specialization

Rule specialization is achieved by detailing other elements of interest for the analyst (for example, web and mobile applications) within rules. If such elements were collected attributes, new rules are generated in the Representation phase. The preparation phase for capturing new data is repeated.

- *Input*: new list of attributes of interest
- *Output*: setup for representing rules with new attributes in the Representation phase or setup for capturing new data in the Capturing phase

Next, we detail how the method has been used to study interactions that involve a group of Facebook users.

5 Representing and Measuring Social Interactions on Facebook

The objective of the experiment presented in this section is to verify the applicability of our method. We represent and measure behavioral contingencies that involve Facebook users in social interactions. Such rules are specialized with media types, and next media-based social interactions are specialized with web applications and devices.

In this paper, we use a data set collected between May 2011 and August 2011 and compare the results with new results obtained from a data set collected between September 2011 and November 2011. Some information about each collected data set is summarized in Table 2. The comparison of the results shows that our method remains consistent over time and generalizes the application of our technique for capturing, representing, and measuring collective actions around media types in social media environments.

5.1 Data Capturing and Preparation

We implemented a Facebook crawler using a Python[1] script, and we ran it between May 2011 and November 2011. We extracted data from more than 1,400 profiles[2] whose owners gave us authorized[3] access.

We collected data about post type (checkin, photo, status, video, link, swf, music, etc.), about user activity (e.g., the addition of *comments* **or** *likes* to a post, the number of users that add both *comments* **and** *likes* to a post, the web application and the mobile device which provides the media object used in social interactions).

---

[1] www.python.org

[2] Most users (954) are from Brazil. Other users come from a variety of countries, such as USA, Canada, Mexico, Argentina, Uruguay, Colombia, England, Portugal, Spain, France, Italy, Belgium, Holland, Russia, Czech Republic, Kosovo, Israel, Turkey, Australia, New Zealand, among others.

[3] We have built a separate Facebook network in which each of the users have both accepted friendship and explicitly authorized the use of information associated with their social interactions.

Table 2 Summarization of information about datasets collected

Dataset	Time of collection	# of users	# of contingencies
OBC 1	May/2011 and August/2011	1,398	102,688
OBC 2	September/2011 and November/2011	1,423	212,138

5.2 Representing and Measuring Social Interactions

In Facebook, a social interaction starts when a user posts on his or on a friend's wall, i.e., a user provides a social stimulus for the social interaction to start. We represent the user who provides the social stimulus as user a.

When user a and the group of his friends f are notified of this post, the group of users k performs the action comment *on that post* and/or users m perform the action like *in a* comment *on that post* and/or users in group l perform the action like *that post*.

In addition, $\bar{k}$ represents the group of users that does not comment *on that post*, $\bar{l}$ represents the group of users that does not like *that post*.

Through observing users' activities on Facebook, the following actions and consequences have been identified for representing social interactions:

- $A_1 =$ post on the wall
- $A_2 = comment$ on that post
- $A_3 = like$ that post
- $A_4 = like$ in a *comment* on that post
- $C_1 =$ be notified of a post (social stimulus)
- $C_2 =$ be notified of *comment*(s) on a post
- $C_3 =$ be notified of *like*(s) for a post
- $C_4 =$ be notified of *like*(s) in a *comment* on a post

Given the set of actions $= \{A_1, A_2, A_3, A_4\}$, users $= \{a, k, l, m\}$ and consequences $= \{C_1, C_2, C_3, C_4\}$ extracted from observing social interactions on Facebook, we have represented these social interactions as behavioral contingencies in Listing 1:

Listing 1

Behavioral Contingencies on Facebook

R1. $aA_1 \cap \bar{k}A_2 \cap \bar{l}A_3 \rightarrow aklC_1$
R2. $aA_1 \cap kA_2 \cap \bar{l}A_3 \rightarrow aklC_1 \cap aklC_2$
R3. $aA_1 \cap \bar{k}A_2 \cap lA_3 \rightarrow aklC_1 \cap aklC_3$
R4. $aA_1 \cap kA_2 \cap lA_3 \rightarrow aklC_1 \cap aklC_2 \cap aklC_3$
R5. $aA_1 \cap kA_2 \cap \bar{l}A_3 \cap mA_4 \rightarrow aklmC_1 \cap aklmC_2 \cap aklmC_4$
R6. $aA_1 \cap kA_2 \cap lA_3 \cap mA_4 \rightarrow aklmC_1 \cap aklmC_2 \cap aklmC_3 \cap aklmC_4$

The implications of Listing 1 are described in the form of *if-then* rules:

R1 *if* user a performs action A_1, users in group k do not perform action A_2, and users in group l do not perform action A_3, *then* user a and users in groups k and l (only) receive consequence C_1 (i.e., user a provides a social stimulus which does not receive any *comments* or *likes*).

R2 *if* user a performs action A_1, users in group k perform action A_2, and users in group l do not perform action A_3, *then* user a and users in groups k and l receive consequences C_1 and C_2 (i.e., user a provides a social stimulus that only receives *comments*).

R3 *if* user a performs action A_1, users in group k do not perform action A_2, and users in group l perform action A_3, *then* user a and users in groups k and l receive consequences C_1 and C_3 (i.e., user a provides a social stimulus that only receives *likes*).

R4 *if* user a performs action A_1, users in group k perform action A_2, and users in group l perform action A_3, *then* user a and users in groups k and l receive consequences C_1 and C_2 and C_3 (i.e., user a provides a social stimulus that receives both *comments* and *likes*).

R5 *if* user a performs action A_1, users in group k perform action A_2, users in group l do not perform action A_3 and users m perform action A_4, *then* user a and his friends (users l, k and m) receive consequences C_1, C_2 and C_4.

R6 *if* user a performs action A_1, users in group k perform action A_2, users in group l do not perform action A_3 and users m perform action A_4, *then* user a and his friends (users l, k and m) receive consequences C_1, C_2, C_3 and C_4.

Figure 3 presents a comparison between the measures of support and cosine correlation for the evaluation using the data sets OBC 1 and OBC 2. The value of *ConR* is 100 % and the *LevR* are 0 %. It must be observed that for each social interaction, if the support (frequency of occurrence) increases, the cosine correlation increases, and if the support decreases, the cosine correlation decreases too. In other words, the increases (or decreases) of frequency of occurrence and the strength (or lack of strength) of association between *B and H* are directly related.

After ranking the rules from maximum to minimum support and cosine correlation levels, we have the rank $R4$, $R1$, $R3$, $R2$, $R5$, and $R6$. This indicates that Facebook users are more engaged in social interactions where the social stimuli receive *comments and likes*. So users are engaged in social interactions where social stimuli do not receive *comments* and do not receive *likes*. Next, users are engaged in social interactions where the social stimuli only receive *likes*. Finally, users are engaged in social interactions where the social stimuli only receive *comments*.

Next, we specialize the social interactions $R4$ (the social interaction which engage Facebook users the most) linking actions performed by users to media types (content of post).

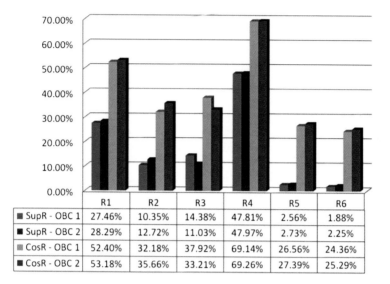

	R1	R2	R3	R4	R5	R6
■ SupR - OBC 1	27.46%	10.35%	14.38%	47.81%	2.56%	1.88%
■ SupR - OBC 2	28.29%	12.72%	11.03%	47.97%	2.73%	2.25%
▨ CosR - OBC 1	52.40%	32.18%	37.92%	69.14%	26.56%	24.36%
■ CosR - OBC 2	53.18%	35.66%	33.21%	69.26%	27.39%	25.29%

Fig. 3 Support and cosine correlation of social interactions

5.3 Media Types Within Social Interactions

On the Facebook mural, users can post media by using text messages, web links, and other media objects. The type of post is automatically identified. The *link* type is used to identify Web links posted on the Facebook Wall via copy/paste—even if it is a Web link for a video, music, photo, etc., from a nonidentified server. The *status* type is used to identify text messages done by the user. The *video* type is used to identify videos shared by users either directly posted on their Facebook Wall, or shared from a Web content provider using an explicit link to Facebook.

The *photo* type is used to identify photos posted by users. The *swf* type identifies applications (generally, animations) in Flash format. The *music* type is used to identify music files posted from a Web content provider. The *checkin* type is the newest type of media (from 1 September 2011) which allows users to confirm their presence in physical locations.

> **Listing 2**
>
> Social Interaction R4 specialized in $aA_1.media$
>
> R4.1. $aA_1.link \cap kA_2 \cap lA_3 \rightarrow aklC_1 \cap aklC_2 \cap aklC_3$
> R4.2. $aA_1.status \cap kA_2 \cap lA_3 \rightarrow aklC_1 \cap aklC_2 \cap aklC_3$
> R4.3. $aA_1.video \cap kA_2 \cap lA_3 \rightarrow aklC_1 \cap aklC_2 \cap aklC_3$
> R4.4. $aA_1.photo \cap kA_2 \cap lA_3 \rightarrow aklC_1 \cap aklC_2 \cap aklC_3$
> R4.5. $aA_1.swf \cap kA_2 \cap lA_3 \rightarrow aklC_1 \cap aklC_2 \cap aklC_3$
> R4.6. $aA_1.music \cap kA_2 \cap lA_3 \rightarrow aklC_1 \cap aklC_2 \cap aklC_3$
> R4.7. $aA_1.checkin \cap kA_2 \cap lA_3 \rightarrow aklC_1 \cap aklC_2 \cap aklC_3$

Through the representation of media types categorized for Facebook, a set of A_1 specialized with the media types was obtained to identify the social stimulus that starts a social interaction.

The media usage within $R4$ is detailed in Listing 2. For example, rule $R4.1$ is described as *if* user a provides a post type *link* as a social stimulus that receives *comments* from users k and *likes* from users l, *then* user a and its friends perceive C_1, C_2, and C_3.

It must be noted that social interactions started by social stimuli *status* and *photo* are frequently shared types of media, whereas *swf* and *music* are less frequently shared types of media.

Table 3 presents the results of the evaluation of the rules presented in Listing 2 with data sets OBC 1 and OBC 2. It must be observed that rule $R4.7$ is not evaluated because the media *checkin* is not present in set OBC 1. We can rank the rules from maximum to minimum levels of $SupR$, $CosR$, and $LevR$ like $R4.2$, $R4.4$, $R4.3$, $R4.1$, $R4.5$, and $R4.6$. We can also rank the rules for OBS 2 from maximum to minimum levels of $SupR$, $CosR$, and $LevR$ as $R4.2$, $R4.4$, $R4.3$, $R4.1$, $R4.7$, $R4.5$, and $R4.6$.

It must be observed that the *status*, *photo*, *video*, and *link* media types engage Facebook users the most with actions *comments and likes* then other media types.

Next, we present the specialization of media-based social interactions, linking actions performed by users and media types for web applications and mobile devices.

5.4 Web Applications and Mobile Devices Within Media-Based Social Interactions

Facebook offers a variety of web applications and mobile devices for users' access and interaction in the social network. By downloading specific applications for mobile devices (e.g., Facebook for iPad, iPhone, Blackberry, and Android), Facebook users can share media types and perform actions *comment and like*. The action *Share* is not available for all mobile devices.

We detail web applications and mobile devices within A_1 to specialize social interactions. For example, $aA_1.video.Links$ means that user a makes a post type *video* provided via web application *Links*. $A_1.status.iPhone$ means that user a makes a post type *status* provided via *iPhone* device.

Table 3 Contingencies $R4.1$ to $R4.7$ and measures for OBC 1 and OBC 2

Rule Number	Media	SupR OBC 1 (%)	OBC 2 (%)	CosR OBC 1 (%)	OBC 2 (%)	LevR OBC 1 (%)	OBC 2 (%)
R4.1	Link	2.30	2.07	44.19	41.15	2.03	1.82
R4.2	Status	22.62	22.44	77.59	80.37	14.12	14.64
R4.3	Video	5.70	4.77	55.10	56.89	4.95	4.24
R4.4	Photo	11.29	11.92	65.51	65.40	7.09	7.53
R4.5	SWF	0.02	0.03	37.80	60.01	0.02	0.03
R4.6	Music	0.02	0.02	55.71	52.34	0.02	0.02
R4.7	Checkin		1.77		65.47		1.68

Listing 3

Web Applications and Mobile Devices Within Media-Based Social Interactions

R4.10 $aA_1.video.Links \cap kA_2 \cap lA_3 \rightarrow aklC_1 \cap aklC_2 \cap aklC_3$
R4.11 $aA_1.photo.Photos \cap kA_2 \cap lA_3 \rightarrow aklC_1 \cap aklC_2 \cap aklC_3$
R4.12 $aA_1.link.Links \cap kA_2 \cap lA_3 \rightarrow aklC_1 \cap aklC_2 \cap aklC_3$
R4.13 $aA_1.status.Twitter \cap kA_2 \cap lA_3 \rightarrow aklC_1 \cap aklC_2 \cap aklC_3$
R4.14 $aA_1.status.iPhone \cap kA_2 \cap lA_3 \rightarrow aklC_1 \cap aklC_2 \cap aklC_3$
R4.15 $aA_1.photo.iPhone \cap kA_2 \cap lA_3 \rightarrow aklC_1 \cap aklC_2 \cap aklC_3$
R4.16 $aA_1.status.Mobile \cap kA_2 \cap lA_3 \rightarrow aklC_1 \cap aklC_2 \cap aklC_3$
R4.17 $aA_1.status.Blackberry \cap kA_2 \cap lA_3 \rightarrow aklC_1 \cap aklC_2 \cap aklC_3$
R4.18 $aA_1.status.Android \cap kA_2 \cap lA_3 \rightarrow aklC_1 \cap aklC_2 \cap aklC_3$

Table 4 presents the results of the evaluation of the rules presented in Listing 3 with the data sets OBC 1 and OBC 2. It must be observed that the media object types *video* and *photo* provided, respectively, via web applications *Links* and *Photos* ($R4.10$ and $R4.11$) engage Facebook users the most with the actions *comment and like*. Also, the *checkin*, *status*, and *photo* media types provided via mobile device *iPhone* ($R4.14$ and $R4.15$) engage Facebook users the most with actions *comment and like*.

5.5 Summarization of Results

By applying our method, we are able to identify that Facebook users engage in social interactions by performing actions *comment and like*. Considering media-based social interactions, the *status* and *photo* types engage Facebook users more by performing actions *comment and like* than other media types.

Table 4 Measures for contingencies R4.10 to R4.22

Rule number	Media and applications	SupR		CosR		LevR	
		OBC 1 (%)	OBC 2 (%)	OBC 1 (%)	OBC 2 (%)	OBC 1 (%)	OBC 2 (%)
R4.10	Video, Links	3.77	1.07	63.16	64.81	3.42	1.05
R4.11	Photo, Photos	2.50	0.97	58.01	62.13	2.31	0.95
R4.12	Link, Links	0.86	0.27	50	50.44	0.84	0.27
R4.13	Status, Twitter	1.12	0.35	60.37	60.14	1.08	0.35
R4.14	Status, iPhone	9.17	2.38	80.60	83.98	7.88	2.30
R4.15	Photo, iPhone	7.48	2.16	79.35	80.31	6.60	2.09
R4.16	Status, Mobile	4.51	1.36	76.98	80.10	4.16	1.33
R4.17	Status, BlackBerry	1.76	0.62	75.94	82.91	1.70	0.62
R4.18	Status, Android	1.49	0.52	74.08	75.95	1.45	0.51

Next, we were able to identify that media types *video* and *photo* provided, respectively, via web applications *Links* and *Photos* engage Facebook users the most with actions *comment and like*. Media types *status* and *photo* provided via mobile device *iPhone* also engage Facebook users the most with the actions *comment and like*.

Application designers may use the results of this analysis to evaluate the sociability of the applications they designed, both as a complementary way to the use of inspection methods of sociability and/or as a model to facilitate the high level of human interpretation of the dynamics of the interaction. This type of analysis is as facilitated by a human-readable model as the one we propose. The method we present in this paper can also be useful to social scientists who want to analyze social interactions, e.g., to investigate the evolution of user behavior in social multimedia environments.

6   Final Remarks

In this chapter, we present a method for capturing, representing, and measuring collective actions around media types as a generalization of applying our human-readable technique used in previous studies. In regard to validating the method, we detail how it has been applied to represent and measure social interactions among a group of 1,600 Facebook users during 7 months.

We verify the application of our method in the representation and measurement of behavioral contingencies that involve Facebook users in social interactions. Such rules are specialized with media types, and next, media-based social interactions are specialized with web applications and devices. Our results report the use of actions *comment and like* after the sharing of *video* and *photo*, respectively, via *Links* and *Photos* web applications and after the sharing of *status* and *photo* provided via the mobile device *iPhone*.

In future works, we plan to apply our method for representing and measuring media-based social interactions in Twitter and Google+ as a follow-up. We also plan to specify a process for software development. By extending the method, we are developing mining algorithms to extract social interactions automatically, including media-based web applications and device-specialized social interactions. Also, we are developing a rule-based model to predict users' behaviors in social media environments.

Acknowledgments We thank CAPES, CNPq, FAPESP, MCT, and FINEP. The author Alan Keller Gomes also thanks PIQS—IFG.

References

Abrol S, Khan L (2010) Tweethood: agglomerative clustering on fuzzy k-closest friends with variable depth for location mining. In: IEEE international conference on social computing (SocialCom). Minneapolis, MN, pp 153–160

Azevedo PJ, Jorge AM (2007) Comparing rule measures for predictive association rules. In: ACM European conference on machine learning (ECML '07). Berlin, pp 510–517

Backstrom L, Bakshy E, Kleinberg J, Lento T, Rosenn I (2011) Center of attention: how Facebook users allocate attention across friends. In: Proceedings of the international AAAI conference on Weblogs and social media

Bentley F, Metcalf C (2009) The use of mobile social presence. Pervasive Comp IEEE 8(4):35–41

Bigonha CAS, Cardoso TNC, Moro MM, Almeida VAF, Gonçalves MA (2010) Detecting evangelists and detractors on twitter. In: Proceedings of 16th Brazilian symposium on multimedia and Web (WebMedia'10), pp 107–114

Bonchi F, Castillo C, Gionis A, Jaimes A (2011) Social network analysis and mining for business applications. ACM Trans Intell Syst Technol 2:22:1–22:37

Chorianopoulos K (2010) Scenarios of use for sociable mobile TV. In: Marcus A, Sala R, Roibs AC (eds) Mobile TV: customizing content and experience. Springer, London, pp 243–254

Cowan LG, Weibel N, Pina LR, Hollan JD, Griswold WG (2011) Ubiquitous sketching for social media. In: Proceedings of the 13th international conference on human computer interaction with mobile devices and services. ACM, New York, NY, pp 395–404

De Choudhury M, Counts S, Czerwinski M (2011) Identifying relevant social media content: leveraging information diversity and user cognition. In: Proceedings of the 22nd ACM conference on hypertext and hypermedia. ACM, New York, NY, pp 161–170

Ekenel H, Semela T (2011) Multimodal genre classification of TV programs and Youtube videos. Multimedia Tools Appl 63:547–567. doi:10.1007/s11042-011-0923-x

Fagá R Jr, Motti VG, Cattelan RG, Teixeira CAC, Pimentel MGC (2010) A social approach to authoring media annotations. In: ACM symposium on document engineering (DocEng '10). ACM, New York, pp 17–26

Freeman LC (2004) The development of social network analysis: a study in the sociology of science. Empirical Press, Vancouver, CA

Fürnkranz J, Gamberger D, Lavrac N (2011) Rule learning: essentials of machine learning and relational data mining, 1st edn. Springer, Dordrecht

Geerts D (2010) The sociability of mobile TV. In: Marcus A, Sala R, Roibs AC (eds) Mobile TV: customizing content and experience. Springer, London, pp 25–28

Geerts D, Grooff DD (2009). Supporting the social uses of television: sociability heuristics for social TV. In: Proceedings of the 27th international conference on human factors in computing systems, CHI 2009. ACM, New York, pp 595–604

Gomes AK, Pimentel MdGC (2011a) Measuring media-based social interactions provided by smartphone applications in social networks. In: Proceedings of the 2011 ACM workshop on social and behavioral networked media access (SBNMA '11). ACM, New York, NY, pp 59–64

Gomes AK, Pimentel MdGC (2011b) Measuring synchronous and asynchronous sharing of collaborative annotations sessions on ubi-videos as social interactions. In: Proceedings of the 2011 international conference on Ubi-media computing (U-MEDIA '11). Sao Paulo, pp 122–129

Gomes AK, Pimentel MdGC (2011c) Social interactions representation as users behavioral contingencies and evaluation in social networks. In: Proceedings of the 2011 I.E. international conference on semantic computing (ICSC '11). Palo Alto, CA, pp 275–278

Gomes AK, Pimentel MdGC (2011d) A technique for human-readable representation and evaluation of media-based social interactions in social networks. In: Proceedings of the 17th Brazilian symposium on multimedia and the Web (WebMedia '11), pp 119–126

Gomes AK, Pimentel MdGC (2011e) Um método para análise de interações sociais na web social como regras se-então (in portuguese). In: Workshop sobre Aspectos da Interação Humano-Computador para a Web Social (WAIHCWS '11) no X Simpósio Brasileiro de Fatores Humanos em Sistemas de Computação (IHC '11)

Gomes AK, Pimentel MGC (2012a) Measuring media-based social interactions in online civic mobilization against corruption in Brazil. In: Proceedings of the 18th Brazilian symposium on multimedia and the Web (WebMedia '12). ACM, New York

Gomes AK, Pimentel MGC (2012b) A media-based social interactions analysis procedure. In: Proceedings of the 27th annual ACM symposium on applied computing (SAC '12). ACM, New York, NY, pp 1018–1024

Gomes AK, Pedrosa DdC, Pimentel MdGC (2011) Evaluating asynchronous sharing of links and annotation sessions as social interactions on internet videos. In: Proceedings of the 2011 IEEE/IPSJ international symposium on applications and the internet (SAINT '11). Munich, pp 184–189

Granovetter MS (1973) The strength of weak ties. Am J Sociol 78(6):13–60

Han J, Kamber M (2005) Data mining: concepts and techniques, 2nd edn. Morgan Kaufmann, Amsterdam

Holland PW, Leinhardt S (1970) A method for detecting structure in sociometric data. Am J Sociol 76(3):492–513

Jin X, Gallagher A, Cao L, Luo J, Han J (2010) The wisdom of social multimedia: using flickr for prediction and forecast. In: ACM international conference on multimedia (MM '10). ACM, New York, pp 1235–1244

Kleinberg J (2000) The small-world phenomenon: an algorithm perspective. In: ACM symposium on theory of computing (STOC). ACM, New York, pp 163–170

Lavrac N, Flach PA, Zupan B (1999) Rule evaluation measures: a unifying view. In: International workshop on inductive logic programming (ILP). Springer, London, pp 174–185

Martínez-Ballesteros M, Riquelme JC (2011) Analysis of measures of quantitative association rules. In: Proceedings of the 6th international conference on hybrid artificial intelligent systems - part II. Springer, Berlin, pp 319–326

Mattaini M (1995) Contingency diagrams as teaching tools. J Behav Anal 18:93–98

Mechner F (1959) A notation system for the description of behavioral procedures. J Exp Anal Behav 2:133–150

Mechner F (2008) Behavioral contingency analysis. J Behav Process 78(2):124–144

Mislove A, Marcon M, Gummadi KP, Druschel P, Bhattacharjee B (2007) Measurement and analysis of online social networks. In: ACM conference on internet measurement (ICM). ACM, New York, pp 29–42

Negoescu R-A, Adams B, Phung D, Venkatesh S, Gatica-Perez D (2009) Flickr hypergroups. In: ACM international conference on multimedia (MM '09). ACM, New York, pp 813–816

Rummel R (1976) Social behavior and interaction – chapter 9. In: Sage Publications (ed) Understanding conflict and war – the conflict. Wiley, New York

Scott JP (2000) Social network analysis: a handbook, 2nd edn. Sage, Thousands Oaks, CA

Shi X, Li Y, Yu P (2011) Collective prediction with latent graphs, In: Proceedings of the 20th ACM international conference on information and knowledge management. New York, NY, pp 1127–1136

Skinner BF (1953) Science and human behavior. New York Press, New York

Tan P-N, Steinbach M, Kumar V (2005) Introduction to data mining. Addison Wesley, Boston

Tan S, Bu J, Chen C, Xu B, Wang C, He X (2011) Using rich social media information for music recommendation via hypergraph model. In: ACM transactions on multimedia computing, communications, and applications (TOMCCAP), 7S, pp 22:1–22:22

Bourlard H, Vinciarelli A, Pantic M (2009) Social signal processing: survey of an emerging domain. Image Vis Comput 27:1743–1759

Wasserman S, Faust K (1994) Social network analysis: methods and applications. Cambridge University Press, Cambridge

Watts DJ (1999) Small worlds: the dynamics of networks between order and randomness. Princeton University Press, Princeton, NJ

Weingarten K, Mechner F (1966) The contingency as an independent variable of social interaction. In: Verhave T (ed) Readings of experimental analysis of behavior. Appleton Century Crofts, New York, pp 447–459

Wilkinson D, Thelwall M (2011) Researching personal information on the public Web. Soc Sci Comput Rev 29(4):387–401

Wilson C, Boe B, Sala A, Puttaswamy KP, Zhao BY (2009) User interactions in social networks and their implications. In: ACM European conference on computer systems. ACM, New York, pp 205–218

Part II
Applications

The Studies of Blogs and Online Communities: From Information to Knowledge and Action

Emanuela Todeva and Donka Keskinova

Abstract This research addresses the question of whether the rise of blogs as a rich information source may create new opinion leaders that transform and challenge the traditionally held public views on drugs and European health care. We investigate blogs that discuss issues related to European health care and European pharmaceuticals for a selected 6 month period. In our approach of the blog space, we take a sociological perspective and design a multistage methodology for data collection and data analysis that differs from the traditionally used crawling techniques by computer scientists.

The results reveal that in spite of the high volume of blogs for the investigated period, only a small number are interlinked by mutual referrals. The emerging network configuration is represented by a small core component with a large number of dyads, or short tails, which represents a fragmented community space. Our content analysis reveals that the information broadcasted in blogs shows emerging semantic differentiation related to specific health issues and disease categories. Our findings support the conclusion that in spite of the high technical Internet connectivity facilitated by search engines and Internet crawling tools, community interaction is limited, and there is no evidence of online crowd or collective action.

1 Introduction

Blogs are related to a group of interactive technologies developed for computer-supported cooperative work in a network and in online virtual environments. These interactive Web 2.0 technologies represent tools for individual online publishing of

E. Todeva (✉)
University of Surrey, Surrey, UK

D. Keskinova
Plovdiv University, Plovdiv, Bulgaria

N. Agarwal et al. (eds.), *Online Collective Action*, Lecture Notes in Social Networks, DOI 10.1007/978-3-7091-1340-0_6, © Springer-Verlag Wien 2014

content, collaborative editing, collaborative learning, or other collaborative online activities. The analysis of these Internet technologies reveals that often the aim of the design (i.e. to facilitate collaboration and interaction) is substituted for the effect of the use of the technology (i.e. the collaboration itself). Such analysis infers social connectivity on the basis of enabled actors, rather than actors actively engaged in interaction. Technological capabilities are substituted with the actual use of the technology and the social processes at the human side of human–computer interaction. The impact of technical connectivity on knowledge creation and dissemination is sometimes analytically substituted for human interaction and community practice.

There is little attempt to discriminate between the use of an interactive technology for broadcasting information and the transformation of this public communication into interaction, information sharing, or a full-scale community collaboration. From a substantive point of view, the question that we address is at what point mass communication becomes internalised in personal lifestyles and when a group of communicating and interacting people becomes a community or when group communication, group awareness, and group coordination transform group interactions into community practices.

In our blog analysis, we aim to discriminate clearly between technical connectivity and human behaviour in an interconnected environment (i.e. interconnected computers and web pages vs. human interactions). Although blogs offer both access to content and a connectivity platform, the presence of technical connectivity should not be used as evidence of a social relationship or other association in an affiliation network. In the first part of this chapter, we develop a theoretical framework for analysis of three distinctive stages of blogging, i.e. information sharing, knowledge creation, and community action and interaction. Our theoretical discussion establishes the conceptual background to our research, looking at the distinction between online communication and online collaboration and interaction.

In order to explore empirically these issues, we build a database of blogs that discuss issues of European pharmaceuticals and European health care for a 6 month period (January–June 2008). Our database includes URL-titles, the referral links between individual blogs, and the semantic context of the blogs. We employ network analysis methods to determine the emergent connectivity.

The methodology for the empirical investigation is outlined in the second part of this chapter, and we label it as a social science methodology for blog analysis. The third part of the paper presents our empirical findings on the structure of the European Pharma blog space, the identification of key blogs that dominate the distribution of information, analysis of the relationship between blogs, evaluation of the semantic structure of the discussion on health issues and disease, and analysis of the referral connectivity in the blog-space.

2   Conceptual Background

Blogs are seen as a new publishing medium—both employed by and challenging the established mass media firms and practices (Gamon et al. 2008). Blogs are portrayed as an enabling technology for personalised expression of opinion and a medium enabling individuals to speak for themselves (Cardon et al. 2007). In all these cases, the expectations are that blogs can and do influence public opinion—either as an Internet broadcasting tool or as a medium used by key opinion leaders, by specialised interest groups, and in communities of practice to promote their activities (Magnus Berquist et al. 2003). As an enabling technology, blogs along with other web technologies facilitate social connectivity on the web, which leads to the evolution of a complex social topology of the web (Barabási et al. 2003).

There are many questions that need to be addressed in this context—both at theoretical and empirical levels. There are also many disciplines that are currently researching the Internet space populated by Web 2.0 users and creators of web content. Research of the web contributes to our knowledge and understanding of the processes that take place in blogging and the impact of various enabling Internet technologies.

Media studies have emphasised that information broadcasting is an effective tool for shaping public opinion and consumer preferences (Eisengerg et al. 1985). At the same time, communication studies confirm that communication relationships require exchange of information, which demonstrates shared meaning between senders and receivers and the creation of a common semantic field. Although communication relationships may take place outside communities, they can take place only within a particular semantic field, where the actors understand each other and share common meaning, semantic frames, and views of the world at large. Semantic fields hence represent the boundaries of distributed communication practices and knowledge and intelligence systems. Broadcasting and receiving information represents a communication relationship that has only one social component—common language code and the meaning of the transmitted content of information. Connectivity within semantic fields, on the other hand, can attribute social connectivity among actors who share a common language, meaning, and interests.

Social connectivity in this sense means social interactions based on the sent and received information and internalisation of this information. Evidence of such social connectivity is any thought or behaviour that is associated with received information. A crowd of people that receive the same information and do the same thing (i.e. broadcast and/or view online information) should be distinguished from coordinated action in a group of people with similar interests (i.e. a community). Social media can involve both broadcasting to a crowd and engaging online community in online information sharing. Both of these trigger collective action, although this is not necessarily equivalent to the received communication.

The literature on online communities has addressed the issues of the nature of virtual communities and the boundaries between communities of practice and knowledge communities as distributed intelligence systems. There is also an effort to draw a distinction between social interactions, human–computer interactions, and online communications (Wellman 2001). Distributed intelligence systems are based on human–computer interactions, information storage and exchange, and online communication. Such a system, however, requires translation of information and social interactions that attribute meaning to the exchanged information in order to resemble a knowledge community within a shared semantic field.

Being part of the Internet, blogs by their nature allow interconnectivity and interaction, and there are expectations that they facilitate the formation of online communities (Cambrosio et al. 2006). The presence of social aggregation (i.e. interconnected social actors), 'long enough' connectivity between these actors, and the enactment of online mediated personal relationships (i.e. interactions and direct one-to-one communication) are seen as evidence of these online communities (Cambrosio et al. 2006). We can assume that online communities that are engaged in repetitive communication share common meaning and represent a knowledge community.

In a community of practice, however, we are looking for evidence of community participation in the 'community activities'. In a distributed intelligence system, there is simply dissemination of information with or without feedback (i.e. with or without interaction). In this context, it is safer to hypothesise communication relationships between blogs and bloggers, rather than community relationships and common patterns of behaviour following community opinion leaders.

Co-authorship and co-editing in distributed intelligence systems are clear evidence of communication relationships as well as evidence of a community membership whereby the co-authors share a common semantic field and act according to community-shared rules and practices. The existing definitions of a community with reference to a group of individuals who share intent, belief, resources, preferences, or needs (Cambrosio et al. 2006) does not allow us to draw clear boundaries—who is a member and who is not a member of a particular community. The sense of belonging is essential (Wellman 2001), and personal statements can be an evidence of a community membership. Link analysis without text analysis can reveal unilateral or reciprocated communication relationships that can substantiate community interactions, but should not be subsumed as such. Feedback and response to communication resemble social interaction that can confirm a level of community affiliation and individual action. Even the most intensive participation in blogs may not be taken as evidence of collective action as it represents communication interaction and not necessarily behavioural contagion.

The typology of blogs circulated in the literature refers to a number of Internet-based shared contents such as online personal diaries, collections of links to other sites, or online public forums devoted to specific topics (Lin et al. 2007). These all are online broadcasting tools that disseminate information and constitute the infrastructure for the online communication process. In terms of their impact, there is a need not only to confirm that the broadcasted information has been viewed by some audience, but also that some form of interaction between a blogger as a sender and a blogger as a recipient of the information has taken place. Clear evidence of impact and social interaction are comments made or acting under the influence of this information. Simple single viewing of information is hardly any evidence of a relationship beyond simple awareness and association. Repetitive viewing or following the html suggestions from a blog, however, may be considered as affiliation—similar to the notion of group membership and acting under the influence of the group. This is referred to as 'other directed' or social contagion of behaviour in crowds (Russ 2007). Posting a comment can be considered as evidence of interaction and a 'community' relationship (Cambrosio et al. 2006). 'Html' referrals in blogs also can be interpreted as enactment of a relationship where the blogger who posts the referral link expresses some knowledge and attitude to the referred blog.

In terms of sources and effects of influence, we have to discriminate between a source of information (i.e. online publishing/broadcasting medium), a potential recipient of information (all Internet users that have online access to the source), and an interaction between the source and the recipient or a reaction by the recipient in response specifically to viewing the source (Todeva 2006a).

Communication in the Internet space is evidence of a membership in a distributed intelligence system, but is still not an evidence of a community relationship as it is not an evidence of shared meaning, sociability, and sense of belonging (Wellman 2001). In spite of the universal connectivity of the Internet, there is a difference between sending and receiving information, shared knowledge and meaning in a common semantic field (i.e. knowledge community), and two or more individuals acting in accord and agreement (i.e. community of practice). An observer to a community differs from a member of that community by the participation in coordinated/shared activities. Community membership for an aggregation of individuals can be attributed by 'sharing' particular beliefs and by enactment of these beliefs in coordinated practices.

The blogger who establishes the blog creates a 'community platform' where other individuals can join-in. A blogger that posts a comment on such a platform demonstrates a clear intent to join this community, and then it is in the realm of text exchanges and meaning sharing, where shared understanding and community relations emerge.

Posting a hyperlink to another blog is merely a unilateral referral to another source of information in a distributed intelligence system and hardly an effective relationship between different blogging communities. Although individual blogs can be interpreted as a community of bloggers, hyperlinks that connect these blogs cannot be treated as evidence of 'a blog community', let alone of an online community of bloggers that participate in discussions in individual blogs. Some evidence of repetitive interactions or impact on behaviour is necessary—to support a 'community' hypothesis for the blogosphere.

In our research project on blogs, we applied an anthropological approach or attempted to collect and analyse information on a selection of blogs that address issues from a specific semantic field (European pharmaceuticals and European health care). We attempted to investigate the 'social' connectivity between these blogs in any form—either by hyperlinks (i.e. distributed intelligence systems) or by participation in online discussions (i.e. knowledge communities and communities of practice). In addition, we looked at semantic connectivity or dominant semantic relationships based on a selection of keywords that represent our semantic framework. Our approach to blog analysis and the methodology for the empirical investigation, described in the next section, have been informed mainly by social anthropology, communication theory, semiotics, and organisation theory.

3 Methodology and Selection Criteria

Blog analysis at present is known as a method for data mining, where the main question is to identify cascading behaviour and to find patterns, rules, clusters, or outliers in the World Wide Web (WWW) and to speculate on the 'potential' spread of influence across blogs linked by referral URLs (Leskovec et al. 2007). New algorithms for page ranking are among the issues that have attracted the attention of the computer scientists (Tseng et al. 2005; Esmaili et al. 2006; Kritikopoulos et al. 2007). Searching core social structures and cyber-communities on the web has led to the development of a number of mapping techniques identifying homogeneous groups of blogs by topic (Dourisboure et al. 2008). More comprehensive analysis of blog content and behaviour has been offered in the context of specific issues such as the political discourse around the American elections in 2004 (Adamic and Glance 2005) or music blogs (Cambrosio et al. 2006).

Our selection of the semantic framework identifies a segment of the blogosphere, which is characterised by an overlap of commercial and public interest. As such, it is expected that the content of the blogs will reflect both commercial and private views. Being a broadcasting media, it is expected also that the content will reflect current events and some deeper underlying individual predispositions to health care and to the pharma industry and products as well as self-expression and sharing of personal experience.

We have chosen to work with the full population of relevant blog URLs over a fixed period of time (January–June 2008) and with a thematic selection of blogs (blogs that have made a reference to at least one of our keywords identified as representative of our semantic framework.

Blog analysis has been associated with blog search and web mining, where the data comes in three main types: content (text, images, etc.), structure (hyperlinks), and usage (navigation, queries, page ranking, etc.). Blog analysis from a computer science perspective implies different techniques such as text, graph, or sequence mining (www 2008). We differentiate from these approaches by developing an alternative methodology for blog analysis that employs simultaneously content and relational analysis in order to evaluate the blog impact as emerging associations in a specified semantic field.

We have adopted a broad agency approach to the Internet where actors can be either pharmaceutical firms discussed in blogs or critical health care issues (expressed by keywords in content) or the blog URL pages themselves. Our application of network analysis of heterogeneous networks aims to reveal the underlying structure of associations between different types of actors that can be interpreted as part of an emergent communication structure in a distributed intelligence system or a knowledge community that shapes a common semantic field and initiates a community action.

We attempted to infer influence by looking at the network position of individual actors in various one-mode and two-mode graphs, where individual actor position is an expression of the set of dyadic relationships of that actor (micro level), the set of relationships within the neighbourhood (mezzo level), and the set of relationships in the entire selected population (global macro level) (Cambrosio et al. 2006).

Our methodology for blog search and blog analysis comprises seven main stages, including building a comprehensive database with the full population of blogs that correspond to our selection criteria, cleaning of the database, evaluation of reference links and semantic associations within blog URLs, and mapping of semantic links and connectivity ties at micro (within blogs) and macro (across blogs) levels.

3.1 Development of the Selection Criteria

Our first step was to demarcate the boundaries of our semantic framework in order to explore it in detail. We conducted text analysis of the news on European health care and European pharmaceuticals broadcasted in official online media between January and June 2008 and identified the 'search keywords' reflecting key events and dominant issues during this period. We grouped these selection words in six distinctive semantic groups (health, drugs, diseases, industry, regulation, region). These groups of keywords demarcated the boundaries of our semantic framework within which we looked at blog referrals and semantic associations within blogs or dyads of keywords with relatively high co-presence in a URL page from the entire blog space.

3.2 Selecting a Blog Search Engine

From a range of blog search engines, we selected Google Blog (http://blogsearch.google.com). The main justification for this decision was that it produced at the time of investigation the minimum duplications of URL pages on initial search, with the maximum of total URL pages identified in a filtered query.

3.3 Search String

We formulated search queries that combined positive and negative filters with Boolean operators such as AND and OR. The positive filter contained three components:

- The scope of the research (pharmaceutical/health care)
- Geophysical relevance (Europe, UK/England, France, Germany, Spain)
- One of the selected keywords (we constrained the sample of keywords for the search with a name of a pharmaceutical company—Pfizer, Glaxo Smith Kline, Sanofi Aventis, Novartis, Hoffmann La Roche, Astra Zeneca, Johnson & Johnson, Merck & Co., Wyeth, Eli Lilly, Bayer, Lacer, Bristol Myers Squibb, Shire Pharmaceuticals, Chiron Corporation, Chugai, Takeda, Teva Pharmaceuticals, Ranbaxy). The use of a company name individualised our queries and enabled us to build a comprehensive database that has entries, which mentioned at least one company name, and the database itself has minimum semantic noise across the population of blogs.

3.4 Building the Database

The final database was generated as the total population of all obtainable blogs that were active at the time of the research and present on the internet for the selected period of 6 months and contained at least one combination of keywords from our search criteria. We downloaded full blog details of these blog URL pages.

3.5 Cleaning the Database

After filtering the majority of duplications by the search engine itself (the difference between visible and obtainable), blogs led to an automated reduction of 85 %, and we cleaned further the database at three additional stages (see Table 1). For this purpose, we used observation techniques and formal techniques based on proprietary software for URL searches. The cleaning of the database passed through the following stages (1) cleaning of duplicate URL pages; (2) cleaning of 'empty URL pages' with size <2 kb information; (3) cleaning of 'shell URL pages' that contain dictionaries, job announcements, lists of URLs without text, URL classifications, and adverts (see Table 1).

According to this procedure, we built a database with the full population of blogs that corresponded to our selection criteria, containing 990 entries. Out of this population, we identified 358 blogs (or 36 %) as an interconnected core that contained shared blogs referring to more than one company (keyword) and 633 blogs (or 64 %) as periphery—i.e. blogs related to only one company (keyword) (observed as pendants on Net 1).

Table 1 Population size

	Total available URLs	Total obtained	Less duplicates and 'shells'[a] (final population)	Representative pages
Total URLs	11,824	2,995	990	633

[a]Duplicate pages—URLs with the same size and content, and registered as different web-links. Shells—URLs that contain lists of words and/or URLs, without other meaningful content

3.6 Developing Blog Attributes (Primary Analysis)

Our primary analysis involved some preliminary observations and developing blog attributes through Internet count of key blog indicators. We calculated four additional indicators as blog attributes per each URL: *size of URL in kb, cross reference between URLs in DB* (as internal hyperlinks), *cross reference to other blogs* (number of external hyperlinks), and *number of occurrences of individual keywords per URL page* (based on the six identified semantic groups). Some of these indicators were used for additional filtering of the data, and the final numbers were recorded after the cleaning process was completed.

3.7 Data Analysis and Network Mapping

The secondary analysis of mapping the selected population of blogs involved mapping of commercial and private broadcasting of information by bloggers and the discussions that this information triggers. We focused on different types of actors that formed a heterogeneous information system. These actors were (1) the owners of the blogs (expressing opinion), (2) the blogs themselves (as cumulative content), and (3) the shared content (through comments, referrals, and embedded URLs). Our network mapping aimed to reveal (1) the emergent structural concentrations of different types of actors and their network position inferring potential information sharing and source of influence, (2) mapping of the content of the text and discussion of information and the semantic links between blogs using one-mode and two-mode graphs, and (3) mapping the emergence of a shared semantic field where we can observe evidence of interactions.

The analysis of these different types of agents required a tool that can deal with heterogeneous systems of actors, and we have selected the approach for heterogeneous relational analysis (Cambrosio et al. 2006). This method allows to analyse the co-occurrence of relationships in a large data set at a dyadic level, which is not available using multidimensional scaling or other clustering techniques.

For the network analysis of the heterogeneous network system, we constituted and interpreted the following 'relationships':

- Associations between blogs and pharmaceutical firms (Nets 1 and 2). The similarity measure in these graphs represents a co-occurrence of pharmaceutical firms in blog contents of individual blogs.
- Association between pharmaceutical firms and semantic categories in five groups (Health, Drugs, Disease, Industry, and Regulation). The similarity measures in these graphs represent significant ties based on a co-occurrence of keywords and names of pharmaceutical companies.
- Associations within semantic fields (Net 3 for Health and Net 4 for Disease). The similarity measures in these graphs represent a co-occurrence of keywords from the semantic groups on health and disease in blog contents of individual blogs.
- Associations and cross reference between blogs (internal links between URL pages in a database) (Net 5). The similarity measure in this graph represents cross references between blogs.

For Nets 1, 2, and 5, we have used the absolute value of ties, and for Nets 3 and 4, we have used values based on standardised residuals (Chi-square) (Todeva 2006b).

4   Overview of Results

The results from the blog analysis are grouped in three main sections (1) mapping of the blog space of our selected framework and mapping the key actors in this space as well as pathways of information dissemination; (2) mapping of the topics on which blog participants publish content (semantic analysis of emerging associations within a knowledge community); (3) mapping of relationships between blogs—comments, feedback, group communications, and community interactions.

4.1   Mapping of the Information Dissemination within the Blog Space

The first two maps show a distribution of URL pages and their association with a particular pharmaceutical company. Net 1 shows that Pfizer, Bayer, and Novartis exhibit unique profiles with their clouds of pendants or URLs dedicated to one company only. These three companies along with two others (Sanofi Aventis and GlaxoSmithKline) form a group of similar firms that are the most referred to in the blog space. These firms have the largest numbers of unique URL pages referring only to one of them (between 50 and 96 referral URLs) and the largest number of URLs that compare two or more of them (between 80 and 183 shared pages for each company). Unique URLs that refer to one company only are graphically presented as pendants on Net 1, and we have labelled them '*representative blogs*', while URLs that refer to two or more firms are labelled '*comparative blogs*'.

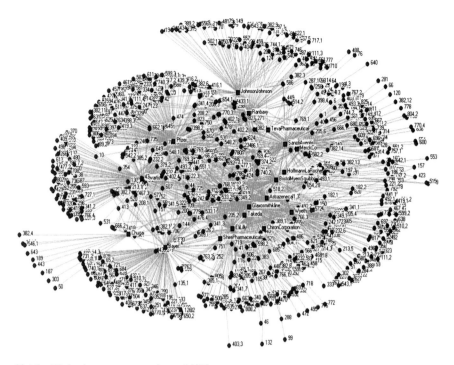

Net 1 All ties between companies and URL-pages

Representative blogs deliver information to the audience regarding one firm only, while comparative blogs enable bloggers to compare and contrast two or more firms.

Among the comparative blogs, we can also discriminate between dyadic and multilateral comparisons that enable bloggers to compare and contrast information on multiple firms. Using the degree centrality measures (DC), we can identify that three of the companies in this group stand out as the most discussed companies with the most similar referral profiles. These are Pfizer (DC = 611), GlaxoSmithKline (DC = 556), and Novartis (DC = 483). The graph in Net 2 displays that while Pfizer and Novartis maintain strong connections with comparative blogs, GSK has significantly low strong ties with blogs (only four connections with comparative blogs).

The second group of seven firms in Net 1 (Merck & Co., Chugai, Takeda, Teva Pharmaceutical, Hoffmann La Roche, Chiron Corporation, Shire Pharmaceuticals) represents an opposite type to the first one—with a minimum number of single referral pages (between 1 and 11) and a fairly low presence in a comparative and competitive context (between 4 and 50 shared URL pages for each company) (Net 1). These firms have a fairly low profile in our population of blogs, which means that bloggers have received information from significantly low Internet sources.

Net 2 More than five ties between companies and URL-pages *(del pendants)*

The third group of six firms (Ranbaxy, Eli Lilly, Wyeth, Johnson & Johnson, Astra Zeneca, Bristol Myers Squibb) stand between the two already described groups, with representative blogs between 16 and 40 and comparative blogs between 37 and 105). These results confirm that the popularity of firms is associated with both unique promotion strategies from representative blogs and online discussions in a comparative and competitive context through comparative blogs. The information impact hence is equally sensitive to volume and strategic approach. The number of comparative blogs is nearly twice bigger (633 blogs) than the number of representative blogs (357 blogs) (Table 1).

The strongest connections between firms and URL pages are exhibited in Net 2, where we observe 24 blogs (out of 990) that have the most intensive ties across 11 firms (out of 18 firms). The interpretation of this map is that these 24 blogs engage in the most intensive comparisons of firms facilitating semantic associations between health issues, pharma companies, and other disease and drugs semantic categories by their co-presence in a common context. Blogs with the most intensive ties to multiple pharma contexts are Pharmalot, Impactivity blog, and Canada's shame.

From the map on Net 2, we can identify two different groups of firms that are most directly compared. These are Pfizer, Astra Zeneca, and Weyth and Beyer, Takeda, and Novartis. The second group of firms mainly emerges through specialised discussions in a few blogs (i.e. blog Asia) that hold strong connections and comparisons.

In terms of scope and reach of the information dissemination, most blogs in our selection have a large number of URL referrals, where 36 % of our population have between 21 and 50 blog referrals, 25 % have between 51 and 100 referrals, 19 % have between 101 and 200 referrals, and 16 % have more than 201 referrals. These numbers indicate a rich information environment where further text analysis may reveal the context for these comparisons.

4.2 Mapping the <u>Semantic Association</u> and Knowledge Creation in Blogs

The concepts that emerge as core categories in our semantic framework are *patient, hospital, disease,* and *medicine.* Our network analysis of this broad semantic context identifies an additional set of keywords that focus the content of the discussion around *health care, public health, community, risk, regulation,* as well as the two most popular diseases—*cancer* and *diabetes* (Nets 3 and 4).

The key actors in our semantic analysis are references in text to pharmaceutical companies or to selected keywords grouped in the semantic subfields of *Health, Drugs, Disease, Industry,* and *Regulation.* In our semantic mapping, we have searched for the co-presence of keywords in blog content and for semantic associations that reveal context and represent knowledge structures. We undertook a semantic analysis of each semantic subfield independently. We present here the results from the semantic analysis of two semantic subfields—*Health* (Net 3) and *Disease* (Net 4).

In our semantic analysis of the first subfield *Health,* we reveal that Astrazeneca, Merck & Co., and Takeda appear most central to the debates surrounding health care issues, including the categories of *medicine, diagnostics,* and *public health.* Pfizer is mostly associated with generic categories such as *health care, health care system,* and *health policy.* Eli Lily appears as an isolate in this map, which means that it has no preferential associations with any particular issues related to health care, but exhibits equal presence in all related discussions. This can be interpreted as 'broad and/or indiscriminate impact'.

The overall structure of the semantic subfield *Health* reveals that some categories are engaged in complex semantic relationships and represent the semantic core, while others are more peripheral and less influential. The semantic categories of *diagnostics, public health, medicine,* and *health care services* represent the core of this semantic field and are instrumental for the discussions around the pharmaceutical companies. The most dominant categories are *health,* which occurs in 70 % of the URL pages, followed by *medicine* (50 %), *patients* (47 %), and *health care* (44 %).

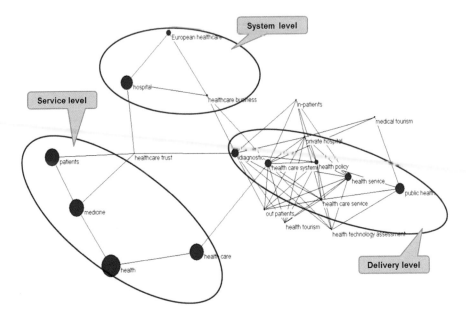

Net 3 Interconnected key words in the semantic field of HEALTH *(normalised value)*

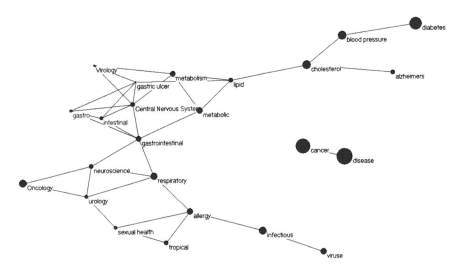

Net 4 Interconnected key words in the semantic field of DISEASE *(normalised value)*

The same semantic subfield *Health* represented in a one-mode graph (Net 3) exhibits different relationships where we observe three distinctive semantic components. These refer to concepts at **service level** (medicine, patients, trust), at **delivery level** (private hospital, outpatients, health tourism, health technology), and at **system level** (European health care, health care business, hospital).

The largest and most densely connected component comprises categories that refer to the delivery system or the restructuring of health care and private health care. The semantic categories interconnected in this component represent associations that underpin the core context of discussions on health. Each of the components contains associations that reflect patterns in the blog content and indicate knowledge structures that require more in-depth text analysis.

The semantic analysis of the subfield *Disease* shows the dominant position of the concept of *cancer* (in 36 % of the URLs) and its association with the generic category of *disease* (in 46 % of the URLs). This semantic association stands alone compared to all other semantic categories. Both concepts of *cancer* and *disease* have the highest occurrence in URL pages (359 and 459 pages respectively) and the highest co-occurrence in the same text. The rest of the categories in the subfield of *Disease* resemble a large interconnected component, bridged by the disease categories of *allergy* (6 % of URLs), *respiratory* (9 %), *gastrointestinal* (5 %), *metabolic* (6 %), and *cholesterol* (13 %). These bridges contain generic disease categories that have a moderate occurrence in the total population of URLs, but have high co-occurrence in the 'disease' subfield (Net 4). All categories in the large component on Net 4 exhibit more rich contexts of the discussions compared with the discussions of *cancer* and *disease*.

4.3 Mapping of the <u>Impact</u> of Information Dissemination and the Emergence of <u>Community Relationships</u>

There are different ways for evaluating the potential impact of an actor. Although communication theory acknowledges that the impact of information dissemination depends on the coding (at the point of the sender—i.e. language, implied meaning, and structure of the message), it also depends on the noise during transmission and the decoding at the point of the receiver. High centrality of senders assures that in their broadcasting role they have high engagement with receivers. The impact of the information, however, has to be sought at the receiver's end.

One of the established methods for evaluation of the impact of blogs and URL pages is how central and interconnected is each URL page. Our analysis in Net 5 reveals very limited impact and connectivity. Our selection of health care blogs shows that out of the 990 URL pages, only 112 are connected, forming 40 disconnected components, 33 of which are dyads, 3 are open triads, and only 4 blogs resemble some form of connectivity with closure and mutual referrals. Most of the blogs are informing only their specific audiences, where the audiences do not seem aware of other blog audiences, i.e. making limited references to other blogs with similar thematic discussion. Most of the external links of our selection of blogs are to URL pages that have more generic content and are not included in our selection.

Nearly 16 % of the blogs in our selection have more than 200 external links or referrals to other blogs, and approximately 20 % have between 101 and 200 external

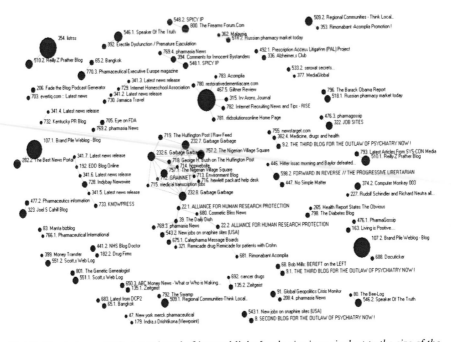

Net 5   Ties between URL-pages based of internal links [*node-size is equivalent to the size of the blog (kb)*]

links. These numbers demonstrate that there are significant efforts to generate internal connectivity, which, however, remains low. The number of dyadic links (33) shows that there are only occasional links between URL pages, and this can be interpreted as evidence of emergent connectivity in the European health care blog space. The majority of blog entries, however, exist mainly by themselves (878 unconnected URL pages). The different size of the nodes in Net 5 is an evidence of some impact associated with a larger volume of information. The blogs with the largest size are Brand Pile, Guiltner Review, and Prather blog.

From the map on Net 5, we can conclude that the blogosphere in our field is very fragmented. There are only occasional links (cross references) between URL pages forming some dyads and short tails among 12 of the blogs in our population (only 12 connected URLs from the total population of 990). There is only one significant component of URL pages in the centre comprising of 21 interconnected URLs linked to a blog called 'Garbage Garbage', where interconnectedness emerges. Such a structural position represents internal 'visibility' for the content in these blogs, where further analysis may reveal the emergence of some interaction patterns. Text and discourse analysis, however, of this interconnected component was beyond our research scope, and, hence, we are not able to draw conclusions on which discussions and semantic associations are linked with this centrality and connectivity.

The final stage of our analysis focused on the question—to what extent the emerging connectivity in the 12 % of the blog space influences the semantic relations and associations within the knowledge field or the impact on the knowledge community of all 990 blogs (Table 1). We compared the results exhibited in Nets 3, 4, and 5 using our measure of internal connectivity (i.e. URL referrals to other blogs internally). We identified that in most cases there are insignificant differences between semantic connectivity within the interconnected 12 % of blogs and the semantic connectivity within the rest of the dispersed population (878 unconnected blogs). The exceptions from this pattern are the categories *diabetes, oncology, cancer,* and *blood pressure,* where the co-occurrence within the knowledge field of interconnected blogs is significantly higher than the use of these concepts in the disconnected population of blogs. The interpretation of these results is that these semantic categories represent stronger public interest that generates connectivity.

Other exceptions are the categories *metabolic* and *respiratory,* where the co-occurrence in the disconnected blogs is significantly higher than in the interconnected core. These are medical categories that have stronger potential impact in their generic use as medical terms rather than in the context of specific semantic associations used by blog users within the emergent knowledge community. Our emergent knowledge community of 112 blogs shares stronger meaning for the first set of categories—*diabetes, oncology, cancer,* and *blood pressure,* where acute public interest generates some form of connectivity in the form of cross references.

5 Conclusions and Managerial Implications

The blog space is a dynamic configuration of the Internet with continuously changing entries and exits. The dynamics is exhibited by a discrepancy between registered new blogs (acknowledged as URL links in the total number of blog searches) and available blogs (blogs obtained for viewing). This is evidence of the noise associated with web crawling and simple computer-based techniques for mapping of the Internet.

The main volume of blogs related to our semantic framework has emerged from the beginning of 2007, and the volume of URL pages has grown rapidly. At the time of this research, it exhibits a dynamic public space where new stories appear continuously, shifting the attention to specific issues. Monitoring this space is essential for tracking major shifts in public opinion, although there is a problem to attribute direct authorship to individual blogs and to measure viewing impact.

Many blogs use automated facilities for organising and structuring the information, e.g. via time-based archiving of posts and tag-based aggregation. These facilities help to establish the semantic frame of individual blogs and facilitate the organisation and distribution of information. The results show emergent semantic frames that can be interpreted as shared communication environments to support online community structure.

There are two main types of blog news—*generalist* news (blogs established by the main media with publications or specialised sections on health care and pharmaceuticals) and *specialised* news (blogs established as specialised sources of information on disease areas, methods of treatment, medicine, and health care). Both types attract fairly similar public attention in terms of comments and interactions, which is still very low (within 1–2 days after the publication). There are also some specific 'community-type' blogs that stir community interactions and some personal blogs—as personal diaries and individual attempts for expression of opinion. We observed very little interaction and public response compared to the volume of text that is broadcasted in publicly accessible blogs. The lack of a significant number of comments can be interpreted as lack of collective action and even as lack of individual action. This is contrary to current observations of oversubscribed social media such as U-Tube and eBay, where we can observe a distinctive volume of individual actions and events of crowd behaviour.

The typology of blogs above suggests that the system for distributed intelligence is not homogeneous and, hence, we may expect different structuring processes to take place. Almost all blogs have included in their registration entry a copyright claim. The majority of blogs have some association with private organisations that manage the blogs, which suggests that serious and long-lasting blogs will exhibit the influence of some institutional strategies and organisational arrangements (McGlohon et al. 2007) that are currently managing this Internet space.

The majority of blog postings represent online broadcasting as they do not receive any comments from the audience, and this excludes the question about emerging online communities. If comments are posted, they often evolve in one thread that follows up upon one article. They were written within 1 to 2 days from the original post and do not engage in a serious discussion or other behavioural response—in order to confirm influence. The majority of blogs in our selected framework can be seen as technical, personal, and organisational experimentations and explorations that aim to broadcast information about drugs, disease, and broad health care issues.

There are many substantially different formats of blogs that are in use, and it seems that there is no dominant pattern of format emerging. Most blogs have options for enabling comments and other interactions such as tagging or emailing an article. However, their classification as blogs and/or their selection by blog search engines is often due to technical features such as meta-tags in the HTML code of the web page.

We have attempted to reveal associations between blogs on the basis of external links (html references to the web pages), internal citation and referencing, and commonality of interests (addressing the same category from our semantic frame).

Due to the high volume of entries in the blog space, research is recommended on a narrow set of categories to demarcate a narrow semantic frame for blog search and for analysis. Our choice of the 19 pharmaceutical companies is a successful strategy as it can draw clear boundaries for the population of URL pages in the database.

For the purpose of our comprehensive analysis, we used two approaches: One included the entire semantic framework, and the other—the leading 19 pharmaceutical firms and keywords in our semantic subfields for health and disease. All pharmaceutical companies included in our search have a presence in the blog space—with the exception of Lacer. Large firms attract a lot more attention, and the reference to Pfizer is dominant (289 URL pages for the period up to July 2008), followed by GlaxoSmithKline (205 URL-pages), Novartis (194), Bayer (159), Sanofi Aventis (156), Eli Lilly (146), and the rest.

The high connectivity between firms in the selected semantic frame indicates that most blogs compare across a wide set of pharma companies, and this may be interpreted as information that is being generated by professional journalists, media activists, or actors with a broad set of observations in the field.

The fragmented internal connectivity between blogs, on the other hand, indicates limited impact from one blog to another blog's audience. Such a fragmented intelligence system cannot offer an effective means for information distribution and its integration as a knowledge community. In addition, the lack of interaction and commenting suggests limited impact and lack of social contagion.

The structural relationships and positions of actors reveal only potential influence, and this is a confirmed observation for all types of actors—for pharma firms, for key blogs, or for semantic categories. In answer to the question '*Could an anonymous blogger have as much influence over public debate as a recognised scientific expert?*', we can answer that although the broadcasting of opinion on the Internet enables access to a global audience, there is no evidence that this audience has been influenced or even engaged at a communication level.

The mapping of the entire blog space for European pharmaceuticals and European health care has a vaguely connected core of 36 % shared URL pages and a large periphery of single URL pages related to a single pharmaceutical company. From the periphery, one third (or 86 URL pages) refer only to Pfizer and to no other pharmaceutical firm from our selection. Although Pfizer appears to occupy a space fairly at a distance from other pharmaceutical firms, it is also directly compared with Johnson & Johnson, Merck & Co., and Takeda, particularly on issues related to *health care*, *health care system*, and *health policy*. In addition, Pfizer appears to be strongly connected to blogs such as Pharmalot, Pharmasia news, RxBlog, Talk: Med, Canada's shame, Computer monkey, Forward in reverse, and Google-Sina medical health—among others. The names of these blogs indicate a very broad information-broadcasting platform that may explain the lack of interaction.

Although Pfizer is a dominant actor in terms of volume of blogs in which it appears in reference, it does not appear to have a distinctive profile. It appears rather generalist—in the semantic subfields of *Drugs, Health,* and *Disease*. This is in contrast to some other pharmaceutical companies that appear closely associated with a particular treatment area or health issues. A more focused discussion can potentially engage bloggers more effectively, particularly if it involves sharing information on personal practice. The lack of such action categories among our semantic categories demonstrates that the online broadcasting of information does not trigger immediate action.

The relationships between URL pages in the blog space are still rare. One of the blogs that has created fairly dense internal and external links is Garbage-garbage, and additional research reveals that few months later it was no longer active. In this context, although we recommend further research on the content of individual blogs with higher centrality, we also acknowledge that these might be texts and information that may disappear from the public space as spontaneously as they had been posted.

The analysis of the semantic subfields of *Health* and *Disease* reveals emerging threads of inter-related issues as well as semantic distances. Among these patterns are close proximity between *medicine, patients,* and *health care trust*—on the one hand—and *private hospitals, medical tourism,* and *health care system*—on the other. These examples indicate emerging discussions in the online public domain.

Disease areas such as cancer, diabetes, and blood pressure are domineering by themselves, and other disease areas such as metabolic, gastrointestinal, and respiratory appear quite interconnected in the public domain. Large pharmaceutical companies appear to have a broader impact on the blog space dominating the context in articles and publications, while small firms appear most often in the shadow of another large pharmaceutical firm. In this context, Pfizer's association with Teva Pharmaceuticals or Wyeth is visible.

The semantic groups of keywords that were identified in our search (health, disease, drugs, industry, and regulation) require independent semantic analysis. Representative research of each semantic subfield is expected to reveal in-depth associations and meaning. Such results will have a direct use in marketing and public relations. Although this semantic analysis reveals connectivity within the blog space, it can be used in analysis of communication flows in other social media.

The unique methodology that we used enabled us to retrieve information on blogs for blog ranking according to their importance in a selected semantic field. Our maps represent contextual graphs that describe locations of URL pages, semantic categories, or firms in context. These maps can be used as guidelines in expertise seeking or finding patterns in blogs' evolution. The unique power of network mapping enables to bridge the gap between micro and macro from individual actors to macro representations of affiliation networks.

Acknowledgements Special thanks and acknowledgement for the contribution of a number of colleagues that actively helped with the finance of the empirical investigation, the development of the methodology, and the analysis of the data (David Parry, Chris Shilling, Hristo Karapchanski, Jana Diesner).

References

Adamic L, Glance N (2005) The political blogosphere and the 2004 U.S. election: Divided they blog, XIVth international world wide web conference, Chiba, Japan
Barabási A, Dezső Z, Ravasz E, Yook S, Oltvai Z (2003) Scale-free and hierarchical structures in complex networks. In: American Institute of Physic Conference Proceedings, 661(1)

Cambrosio A, Keating P, Mogoutov A (2006) Mapping the emergence and development of translational cancer research. Eur J Cancer 42(18):3140–3148

Cardon D, Delaunay-Teterel H, Fluckiger C, Prieur C (2007) Sociological typology of personal blogs, ICWSM'2007. International conference on weblogs and social media, Boulder, CO

Dourisboure Y, Geraci F, Pellegrini M (2008) Extraction and classification of dense communities in the web. XVIth international world wide web conference, Banff, AB

Eisengerg E, Farace R, Monge P, Bettinghaus E, Kurchner-Hawkins R, Miller K, Rothman L (1985) Communication linkages in inter-organisational systems. In: Dervin B, Voight M (eds) Progress in communication sciences, vol 6. Ablex, Norwood, NJ, pp 210–266

Esmaili K, Jamali M, Neshati M, Abolhassani H, Soltan-Zadeh Y (2006) Experiments on persian weblogs. XVth international world wide web conference, Edinburgh

Gamon M, Basu S, Belenko D, Fisher D, Hurst M, Konig A (2008) BLEWS: using blogs to provide context for news articles. Association for the Advancement of Artificial Intelligence

Kritikopoulos A, Sideri M, Varlamis I (2007) Blogrank: ranking on the blogosphere, ICWSM'2007, international conference on weblogs & social media, Boulder, CO

Leskovec J, McGlohon M, Faloutsos C, Glance N, Hurst M (2007) Cascading behaviour in large blog graphs. SIAM Data Mining

Lin J, Halavais A, Zhang B (2007) The blog network in America: blogs as indicators of relationships among US cities. Connections 28(2):22–30

Magnus Berquist M, Feller J, Ljungberg J (eds) (2003) Open source software movements and communities. In: Proceedings of the international conference on communities and technologies, Amsterdam, Netherlands, September, 2003

McGlohon M, Leskovec J, Faloutsos C, Hurst M, Glance N (2007) Finding patterns in blog shapes and blog evolution. ICWSM'2007, International conference on weblogs and social media, Boulder, CO

Russ C (2007) Online crowds—extraordinary mass behavior on the internet. In: Proceedings of I-MEDIA '07 and I-SEMANTICS '07, Graz, Austria, September 5–7, 2007

Todeva E (2006a) Business networks: strategy and structure. Taylor & Francis, New York, NY

Todeva E (2006b) Clusters in the south east of England. University of Surrey, Surrey

Tseng B, Tatemura J, Wu Y (2005) Tomographic clustering to visualize blog communities as mountain views. XIVth international world wide web conference, Chiba, Japan

Wellman B (2001) Computers as social networks. Science 293(5538):2031

www (2008) An introduction to web mining. http://www2008.org/program/program-tutorials-TF3.html

Using Contemporary Collective Action to Understand the Use of Computer-Mediated Communication in Virtual Citizen Science

Jason T. Reed, Arfon Smith, Michael Parish, and Angelique Rickhoff

Abstract Virtual citizen science creates Internet-based projects that involve volunteers who collaborate with scientists in authentic scientific research. Forms of computer-mediated communication like websites, email, and forums are integral for all project activity and interaction between participants. However, the specific forms of computer-mediated communication vary because of the functionalities they must serve for a particular virtual citizen science project. After illustrating how computer-mediated communication is used by the Zooniverse, a collection of successful virtual citizen science projects, this chapter describes how virtual citizen science can be understood as a form of online collective action that takes place in the context of conducting scientific research. Using collective action theory allows for the creation of a collective action space that can be used to compare particular project features or entire projects based on a combination of the forms of interaction and project responsibilities available to volunteers.

1 Introduction

Advances in technology have undoubtedly and dramatically impacted many aspects of contemporary science practices. In particular, ever-evolving forms of computer-mediated communication (CMC) like email, chat, text messaging, and Web-based platforms increase the opportunities for interpersonal interactions among the parties involved in scientific research (Walther 1996). For example, scientists use CMC to create virtual research environments known as collaboratories to aid in long-distance research collaborations with other scientists and researchers (Finholt

J.T. Reed (✉) • A. Smith • M. Parish • A. Rickhoff
Adler Planetarium, Chicago, IL 60605, USA
e-mail: jreed@adlerplanetarium.org; arfon@zooniverse.org; michael@zooniverse.org;
arickhoff@adlerplanetarium.org

N. Agarwal et al. (eds.), *Online Collective Action*, Lecture Notes in Social Networks, DOI 10.1007/978-3-7091-1340-0_7, © Springer-Verlag Wien 2014

2002). The same CMC that connects professional scientists to each other also enables interested members of the public (e.g., hobbyists, amateurs) to directly contact the scientists or other enthusiasts.

An interesting intersection of contact between scientists and the public as part of conducting a scientific investigation is in the practice of citizen science. Like many examples of scientific inquiry, citizen science projects are created and run by professional scientists with expertise in the topic under and methods of investigation. A hallmark of citizen science is that these projects give volunteers, who often are members of the public with no expertise in scientific investigation, the opportunity to become directly involved in using authentic scientific methods in the gathering, analysis, interpretation, and dissemination of data (Cohn 2008). With careful planning and execution, citizen science creates data used to draw reliable and valid scientific conclusions from volunteer efforts (Silvertown 2009). Besides adding to the body of scientific knowledge, citizen science project results can also contribute to civic betterment through avenues like policy change and conservation effort (Irwin 1995).

Like other forms of scientific investigation, citizen science projects have incorporated CMC into their practices. This chapter will provide a detailed look at CMC as part of citizen science. We begin by detailing how technology-mediated participation has become integrated into the general practice of citizen science. We continue with an illustration of these ideas at work with the example of the Zooniverse, a set of online citizen science projects. We conclude with an examination of how online collective action as a form of communication aids in understanding the nature of CMC in citizen science.

2   The Integration of Computer-Mediated Communication into Citizen Science

Understanding how citizen science projects use CMC requires moving beyond an assumption that they are just another example of computer-mediated technology (Wiggins and Crowston 2011). Unlike collaboratories that use CMC primarily to connect the efforts of professional scientists, citizen science uses it to connect research professionals with nonexperts in the public. Furthermore, citizen science is not self-organizing like many open-source or peer production efforts that use CMC to allow the participants to determine and manage the tasks and their execution. Instead, citizen science evidences a more hierarchical structure in which the scientists determine the procedural protocols that volunteers must follow. This clarifies the role of volunteers in the research project so their efforts turn into meaningful contributions by creating useful and useable data.

Of course, the use of CMC varies across citizen science projects. Many citizen science projects are hybrids that use CMC to enhance traditional citizen science practices. One of the most common examples of such hybridization is using CMC to increase the effectiveness of gathering data. For example, placing all project materials for the 2010–2011 Christmas Bird Count on a website run by the National Audubon Society led to a record 62,624 volunteers providing over 61 million observations of birds in over 2,200 different locations (LeBaron 2011). Web-based technology can also enable citizen science projects to maximize the scope of the research topic with a limited amount of resources. The Road Watch in the Pass project created a site that allowed less than 1 % of the total driving population of an area to contribute observations of wildlife near highways that were similar in quality to those of highway contractors responsible for carcass removal (Lee et al. 2006). These examples demonstrate how adding CMC to traditional citizen science practices can increase the amount and quality of volunteer contributions.

A growing class of citizen science projects known as *virtual citizen science* (VCS) moves from hybridization to complete integration of CMC into all project activity. Unlike more traditional citizen science, all project activity and participant interaction in VCS occur via CMC like websites, email, and online forums (Wiggins and Crowston 2011). Importantly, VCS uses the CMC with the most appropriate functionality for project activity and participant interaction (Prestopnik and Crowston 2012); the likelihood of VCS project success is increased if the technology used appropriately supports the goals of the project. For example, VCS relies on attracting sufficient amounts of volunteer activity towards the scientific goals of the project. Websites provide the flexible platform scientists need to create protocols appropriate to transform volunteer activity into meaningful data. The websites also expand the available pool of volunteers by lowering the barriers of access to the VCS project; volunteers only need access to the VCS project website to participate. CMC like forums, chat, and blogs provide communication channels that aid volunteer interaction in VCS. One example of these ideas in action can be found in a collection of successful VCS projects called the Zooniverse.

3 The Zooniverse: An Illustration of Virtual Citizen Science

The first Zooniverse project called Galaxy Zoo was created in 2007 to classify the shapes of individual entries in the Sloan Digital Sky Survey, a 60-terabyte survey of the stars equivalent in size to the entire body of knowledge housed in the Library of Congress (Raddick et al. 2009). Labeling of morphological features of the raw data was necessary before any other research investigations could take place. They also required multiple classifications of each image in the database as a way of assessing their reliability. The sheer amount of data prevented any individual or small group, let alone experts in astronomy, from classifying the raw data. Conveniently, this kind of task can be structured so that it only requires the ability to assess patterns without any formal training in astronomy. VCS is well-suited to this kind of task because technology cannot yet reliably outperform humans in this basic perceptual ability and therefore requires human effort. Thus, the effort of large numbers of individuals simply willing to try is both necessary and sufficient for the successful completion of this task.

The scientists behind the project created a website with a basic interface consisting of images to classify and multiple-choice questions about various morphological features of the images. In response, members of the public contributed about eight million classifications within the first 10 days of the project (Fortson et al. 2012). This classification activity at the launch was so intense that it overwhelmed the single project server and took the Galaxy Zoo Website offline for four of the first hours of its existence. Contribution levels were just as strong when the website went back online. By the end of its first year, about 150,000 people provided more than 50 million classifications (Pinkowski 2010).

This level of success also led to two revisions of Galaxy Zoo. The aptly named Galaxy Zoo 2 expanded the classification task to include more detailed questions. The subsequent version named Galaxy Zoo: Hubble applies the task from Galaxy Zoo 2 to an entirely new set of pictures from the Hubble Space Telescope. Perhaps more important than the impressive response rate or life of the project is that the efforts of the volunteers were collectively just as good as the judgments of expert astronomers. The classifications from the public on the Galaxy Zoo projects directly contributed to the publication of 25 papers in peer-review scientific journals.[1]

Besides providing quality data for scientific research, the virtual platform of Galaxy Zoo also engaged volunteers beyond the data classification task. An unexpected flood of emails from the Galaxy Zoo volunteers to the scientists running the project required the creation of a separate forum section to effectively address the number and scope of these public inquiries. This forum created a direct channel of communication between the participants and enhanced the collaborative nature of this endeavor for the members of the public (Fortson et al. 2012).

In addition to satisfying volunteers' demand for a channel to communicate about their experiences, the Galaxy Zoo forum also led to the serendipitous discovery of two new astronomy phenomena. Hanny van Arkel, a Dutch schoolteacher, used Galaxy Zoo like most volunteers to classify morphological features of galaxies and ask questions on the forum about the pictures she saw during this task. Two of those posts featured questions similar to those she had previously posted on the forum. Hanny (also her username on the forum) is not a professional astronomer, so she began using the forum to seek answers about various features of pictures she had seen during her use of Galaxy Zoo. One of these forum posts titled "Give peas a chance!" showed a picture of a small green object.[2] The next day, Hanny contributed another forum post asking for information about the "blue stuff" in a different picture.[3]

---

[1] http://www.galaxyzoo.org/published_papers

[2] http://www.galaxyzooforum.org/index.php?topic=3802.0

[3] http://www.galaxyzooforum.org/index.php?topic=3638.0

This so-called blue stuff caught the attention of other forum users. Another user, playing off of Hanny's nationality and astronomy conventions of naming objects after their discoverer, suggested that it should be called Hanny's Voorwerp, which is Dutch for "Hanny's object."[4] The name became official as a forum moderator started to use it in future discussions of the object. The official nature of the name became important once the "Zookeepers," professional astronomers who maintained an active presence on the forum, realized that the "blue stuff" was a previously undiscovered class of object (Fortson et al. 2012). Hanny's original question spurred discussion among scientists and volunteers on the forum that culminated in actual research on a new and unexpected discovery (e.g., Lintott et al. 2009; Rampadarath et al. 2010; Schawinski et al. 2010).

Hanny's other post about peas also prompted other volunteers to post to the forum. Unlike the Voorwerp discussion thread that mainly featured postings by the Zookeepers detailing their ongoing investigation of the "blue stuff," the discussion thread here featured other volunteers posting their own examples of the green objects as well as their own puns and plays on words involving peas. In essence, this thread contained elements of both task-related and social discussion. Importantly, this activity by the volunteers provided enough examples for Zookeepers to analyze and confirm the discovery of another new class of object, the "Green Pea" galaxies (Cardamone et al. 2009).

The success of the Galaxy Zoo project prompted the creation of the Zooniverse. The Zooniverse acts a central hub for VCS projects that follow the same basic idea of Galaxy Zoo to provide volunteers with an accessible online project where their contributions are part of a collective effort to create meaningful datasets and engage in authentic scientific research. Because of the success of the original Galaxy Zoo project, subsequent Zooniverse project designs featured both some kind of data classification task as well as some form of forum. Some projects used a task similar to Galaxy Zoo, where volunteers classified objects by picking from a closed set of choices. Other Zooniverse projects provided volunteers with an open-ended task that required them to mark objects of interest on data images (Fortson et al. 2012). By using a form of the task appropriate to the purpose of the investigation, other Zooniverse projects also attracted sufficient amounts of volunteer activity to completely analyze the raw data available. Some Zooniverse projects have added new data for the volunteers to continue working with, whereas others were retired because the projects used all available data.

In addition to keeping the strengths of Galaxy Zoo, subsequent Zooniverse projects evolved new forms and features to enhance the success of projects. For example, volunteers showed more activity in the data classification tasks than posting information in the forums of some of the earlier Zooniverse projects. The Zooniverse development team recognized that more could be done to encourage volunteers to interact with the projects beyond the data classification tasks and created the "Talk" tool. The "Talk" tool first appeared in a Zooniverse project

---

[4] http://www.galaxyzooforum.org/index.php?topic=279585.0

called Planet Hunters[5] that asked volunteers to find new planets by identifying patterns of light changes in stars that could represent a possible planet. Unlike Galaxy Zoo, volunteers in Planet Hunters drew boxes onto the parts of a picture that they believed signified the desired changes in light transmission. After volunteers finished marking an image, they were directly asked whether they would like to discuss their observations.

If a volunteer accepted the invitation, they would go to a screen that presented all options of the "Talk" tool. As with the forums in the previous Zooniverse projects, volunteers could make whatever comments they wished. However, the "Talk" tool provided additional features like more detailed scientific information about the image they just classified, opportunities to mark that image as part of a personal collection of interesting objects, and other discussions about that particular image. The "Talk" tool also directly led to the discovery of candidates for new planets. On January 17, 2012, the Planet Hunters project was featured on a British Broadcasting Corporation program and issued a challenge for viewers to make 250,000 classifications in 48 h. Volunteers contributed over one million classifications in that time period and also spotted possible candidates for new planets (Meg 2012).

These examples from Zooniverse demonstrate how CMC serves various purposes in VCS projects. Each Zooniverse project used a form of task interface appropriate to its data classification needs. The forums and the "Talk" tool were created in response to various issues around the interpersonal communication for each project. Although the success of these choices is apparent, there is more to understand about how and why the success occurred. One answer may lie in the work on online collective action.

4 Online Collective Action and Virtual Citizen Science

VCS can be conceived of as a form on online collective action that takes place in the context of conducting scientific research. It is important to confine our comparison to a domain like scientific research in which all participants are working towards the same end of completing the research project. Wiggins (2012) noted that the purpose of VCS to advance scientific research lacks the political antagonists and opposition more commonly seen in online collective action. Put another way, online collective action is more applicable in studying confrontational contexts than VCS. We fully recognize that online collective action can and does engage in many forms of pro-social action; however, there is also an "us vs. them" aspect more applicable to online collective action than VCS. As such, online collective action is most informative to the study of VCS in cooperative contexts like scientific research.

Theory from online collective action provides an apt description of the technology and actions found in VCS. Wiggins (2012) uses the language of collective

---

[5] www.planethunters.org

action theories in describing VCS projects as providing volunteers a "facilitative opportunity structure" (p. 301); the project websites provide volunteers access to the protocols necessary for their activities to directly and meaningfully contribute to scientific research. In this way, the volunteers contribute to the development of a common good, or some product that anyone may use no matter what their individual contribution. Common goods in virtual citizen science would most likely take the form of the data gathered for a project; for example, volunteers' contributions are aggregated and converted into usable data.

Creation of the common goods of VCS is aided by technology like the Internet increasing the permeability between people's private and public activity (Bimber et al. 2005). Virtual platforms make it easier than ever for an individual to contribute some amount of their effort towards a project, cause, or issue. The size of the contribution notwithstanding, that individual's effort—the private behavior—becomes part of the overall effort towards creating a public good. As long as another person also contributes his or her efforts towards that same public good, collective action has occurred.

Bimber et al. (2005) also explain how blurring the divide between private and public activity creates a new dynamic in contemporary collective action. One of the biggest concerns in traditional models and examples of collective action was to prevent free riding while encouraging participation. However, free riding is only a concern when a person considers a discrete decision of weighing the cost of participating against the benefits of creating a public good. The ease of access and low cost of participation provided by Web-based forms of collective action effectively removes such considerations. The large numbers of volunteers and even larger number of their contributions in VCS projects like Galaxy Zoo are consistent with this idea. The ease of access that these Web-based projects offer is a low cost for the volunteers, thereby greatly reducing if not eliminating considerations beyond how their efforts contributed to the public good of classifying data.

Decreasing the distinction between private and public activity also increases the amount and type of interaction available as part of contemporary collective action. This flexibility aptly describes the data classification activities as well as the forum discussions seen in Zooniverse projects like Galaxy Zoo. Like traditional forms of collective action, the data classification activities use a hierarchy with the scientists creating situations for the volunteers to contribute their efforts towards filling the scientists' goals of gathering usable data. The discovery of Hanny's Voorwerp and the Green Peas galaxies also demonstrate interaction consistent with contemporary collective action. The discussion thread involving the Green Peas featured volunteers posting their own examples of those objects in response to the posts of other volunteers. The volunteers were taking advantage of the opportunity for interaction about topics that mattered to them.

This is perhaps best reflected in what Flanagin et al. (2006) term the collective action space. It is comprised of two orthogonal dimensions. One dimension details the mode of interaction: volunteers can experience no direct interaction with another person on one extreme compared to repeated direct interactions with known others on the other extreme. The other dimension details the mode of engagement: the activities available to volunteers present low-responsibility tasks with little opportunity for pursuing matters important to the self on one extreme compared to tasks with high responsibility and opportunity for pursuing matters important to the self on the other extreme.

This creates a set of four quadrants with which to describe collective activity (see Fig. 1, based on Flanagin et al. 2006). Quadrant I describes situations in which volunteers can engage in tasks that present them with high amounts of responsibility and opportunity but lack direct interaction with other people. Quadrant II also offers high responsibility and opportunity in engagement but with direct interaction with other people. Quadrant III describes situations in which volunteers can engage in tasks that have low responsibility and opportunity to pursue personal matters but feature direct interaction with other people. Finally, Quadrant IV describes situations in which volunteers can engage in tasks that have low responsibility and opportunity to pursue personal matters and feature no direct interaction with other people.

This collective action space is very applicable to understanding the nature of virtual citizen science projects. We have added a diagonal line to Fig. 1 through Quadrants II and IV that represents the activity on the Galaxy Zoo project. The classification of galaxies' activity can be described well by Quadrant IV; Galaxy Zoo volunteers contributed their effort towards the very straightforward task of classifying the shape of what they saw by choosing from a closed set of options. This task also did not require nor necessarily encourage volunteers to seek out interaction with other volunteers or project scientists. On the other end of the dotted line is the example of the "Green Peas" discovery. This is an example of a Quadrant II activity. Unlike the classification of galaxies, the discovery of the "Green Peas" would not have occurred if volunteers lacked the high responsibility and opportunity to post their own example of potential Green Pea pictures. It also relied on high levels of direct interaction with other volunteers, reacting to both the topic of interest (e.g., the photos) and the social aspects (e.g., comments and word play related to the concept of peas).

Figure 1 also places the discovery of Hanny's Voorwerp in Quadrant II but not as high on either dimension as the "Green Peas" discovery. This is to demonstrate that the collective action space can also describe different aspects of the serendipitous discoveries made in the Galaxy Zoo project. Compared to the "Green Peas" forum discussion, the discussion about the Voorwerp involved more discussion between the project scientists. This also reflects how there was comparatively less activity of other volunteers in posting their own observations or comments; they still made contributions of their own thoughts and conversation, but not in the same way as the Green Peas discussion.

Besides being able to provide a framework for understanding how the mode of interaction and mode of engagement vary within VCS projects, this collective action space also allows comparison of these dimensions between VCS projects. Using the collective action space, it is a straightforward matter to compare how projects like Planet Hunters with the "Talk" tool compare to projects like Galaxy Zoo that had the forum separate from the data classification task. The classification

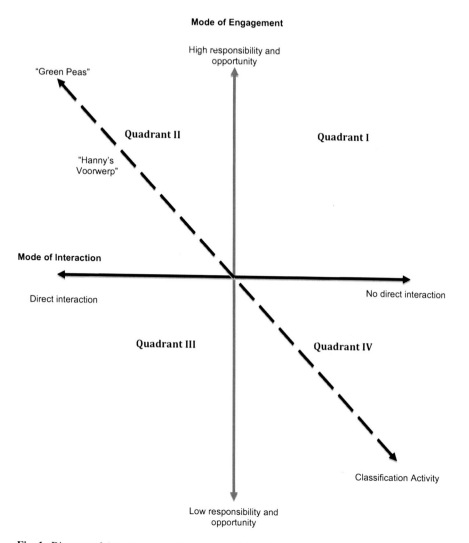

Fig. 1 Diagram of the collective action space and description of the Galaxy Zoo project

of light reflection in Planet Hunters is similar to the Quadrant IV data classification of Galaxy Zoo; both present fairly straightforward tasks that the scientists have designed to provide an authentic but accessible experience with science that volunteers can complete on their own. The "Talk" tool also presents a Quadrant II experience similar to the forum discussions about Hanny's Voorwerp and the Green Peas.

Where these two Zooniverse projects begin to differ is in how the "Talk" tool can increase the chances of projects like Planet Hunters extending their footprint into Quadrants I and III. Because the "Talk" tool is part of the data classification task in Planet Hunters, it is more likely that volunteers may mark a particular image as interesting and add it to their personal collection but not make a comment about it. This would be consistent with a Quadrant I action where the volunteers take advantage of the opportunity presented by the "Talk" tool to note a particularly interesting image and explore more data about it, but decline to share it with other people. Volunteers could also use the "Talk" tool to follow discussions about an object and perhaps make comments, but not mark it as interesting or contribute information of substance. This would be an example of a Quadrant III activity in which volunteers interact with other people, but do not take on any particular responsibility or opportunity. These types of activities would be possible in Galaxy Zoo, but they are more likely to occur in projects like Planet Hunters as a result of the "Talk" tool being directly tied to the data classification task.

5 Conclusion

Approaching virtual citizen science from the perspective of contemporary collective action yields a number of benefits. While recognizing the powerful changes that new forms of technology have had on collective volunteer efforts towards producing a common good, this approach appropriately couches it in terms of the communication and not the technology. In essence, "many of these 'novel' forms are not necessarily new but, rather reflect alternative ways of organizing that have always been available but only now, when technological developments have enabled more opportunities for these forms to emerge and be sustained, have the dynamics become readily apparent" (Flanagin et al. 2006, p. 33). Virtual citizen science can use a similar perspective that focuses on the interaction between volunteers, scientists, and the projects while recognizing the importance, but separateness, of the technology involved.

Contemporary forms of collective action offer guidance in understanding the nature of virtual citizen science. Advances in technology create new forms of communication between volunteers and scientists. This focus provides a collective action space with which to compare virtual citizen science projects based on the nature of the interaction and engagement offered to participants. As the breadth and scope of virtual citizen science projects spread, frameworks like this will offer a guide to researchers and practitioners to make more informed choices about which aspects of virtual citizen science to study and develop.

Projects like those found in the Zooniverse are no exception. The collection of citizen science projects offered by the Zooniverse continues to build on the success of the original projects like Galaxy Zoo and new developments like the "Talk" tool in projects like Planet Hunters. As new VCS projects are added to the Zooniverse, approaches like the collective action space will help inform scientists about what features to implement in their continuing efforts to make VCS as meaningful as possible for both volunteers and scientists.

Acknowledgments This material is based upon work supported by the National Science Foundation under Grant No. 0917608. Any opinions, findings, and conclusions or recommendations expressed in this material are those of the authors and do not necessarily reflect the views of the National Science Foundation.

References

Bimber B, Flanagin AJ, Stohl C (2005) Reconceptualizing collective action in the contemporary media environment. Commun Theory 15:365–388

Cardamone CN, Schawinski K, Sarzi M, Bamford SP, Bennert N, Urry CM, Lintott C et al (2009) Galaxy zoo green peas: discovery of a class of compact extremely star-forming galaxies. Mon Not R Astron Soc 399:18

Cohn JP (2008) Citizen science: can volunteers do real research? Bioscience 58:192

Finholt TA (2002) In: Cronin B (ed) Annual review of information science and technology. American Society for Information Science, Washington, DC, pp 74–107

Flanagin A, Stohl C, Bimber B (2006) Modeling the structure of collective action. Commun Monogr 73:29–54

Fortson L, Masters K, Nichol R, Borne K, Edmondson E, Lintott C, Raddick J, Schwaniski K, Wallin J (2012) Galaxy zoo: morphological classification in citizen science. In: Way JM, Scargle JD, Ali K, Srivastava AN (eds) Advances in machine learning and data mining for astronomy. CRC Press, Boca Raton, pp 213–236

Irwin A (1995) Citizen science: a study of people, expertise, and sustainable development. Environment and society. Routledge, New York

LeBaron GS (2011) The 111th Christmas bird count. Am Birds 65:2–7

Lee T, Quinn MS, Duke D (2006) Citizen, science, highways, and wildlife: using a web-based GIS to engage citizens in collecting wildlife information. Ecol Soc 11:Article11. Retrieved 29 Jan 2012, from http://www.ecologyandsociety.org/vol11/iss1/art11/

Lintott CJ, Schawinski K, Keel W, Van Arkel H, Bennert N, Edmondson E, Thomas D et al (2009) Galaxy zoo: "Hanny's Voorwerp", a quasar light echo? Mon Not R Astron Soc 399 (1):129–140

Meg (2012) Stargazing final results [Web log comment]. Retrieved from http://blogs.zooniverse.org/planethunters/2012/01/18/stargazing-final-results/

Pinkowski J (2010) How to classify a million galaxies in three weeks. Time. Retrieved from http://www.time.com

Prestopnik N, Crowston K (2012) Citizen science system assemblages: toward greater understanding of technologies to support crowdsourced science. Paper presented at the 2012 iConference, Toronto, ON

Raddick MJ, Bracey G, Carney K, Guyek G, Borne K, Wallin J, Jacoby S (2009) Citizen science: status and research directions for the coming decade. ASTRO2010 decadal survey position paper

Rampadarath H, Garrett MA, Józsa GIG, Muxlow T, Oosterloo TA, Paragi Z, Beswick R et al (2010) Hanny's Voorwerp: evidence of AGN activity and a nuclear starburst in the central regions of IC 2497. Astron Astrophys (14782):5. Retrieved from http://arxiv.org/abs/1006.4096

Schawinski K, Evans DA, Virani S, Urry CM, Keel WC, Natarajan P, Lintott CJ et al (2010) The sudden death of the nearest quasar. Astrophys J 724:4

Silvertown J (2009) A new dawn for citizen science. Trends Ecol Evol 24:467–471

Walther JB (1996) Computer-mediated communication: impersonal, interpersonal, and hyperpersonal interaction. Commun Res 23:3–43

Wiggins A (2012) Crowdsourcing scientific work: a comparative study of technologies, processes, and outcomes in citizen science. Unpublished doctoral dissertation. Syracuse University, Syracuse, NY

Wiggins A, Crowston K (2011) From conservation to crowdsourcing: a typology of citizen science. In: Proceedings of the 44th annual Hawaii international conference on system sciences. Koloa, HI

Socially Networked Citizen Science and the Crowd-Sourcing of Pro-Environmental Collective Actions

Janis L. Dickinson and Rhiannon L. Crain

Abstract The social Web has changed the nature of human collaboration with new possibilities for massive-scale cooperation in such important endeavors as scientific research and environmentally important collective action. While first generation citizen science projects have successfully used the Web to crowd-source environmental data collection, "next generation" citizen science practice networks combine crowd-sourcing, joint sense of purpose, and soft institutional governance with the distributed intelligence and efficacy of online social networks. Here we tap into evolutionary theory and social psychology to generate hypotheses for how such "next generation" citizen projects can best support pro-environmental behaviors like habitat restoration and energy conservation. Recent research on the evolution of cooperation highlights the potential for reputational mechanisms and scorekeeping to foster cooperation in online social networks. Nested bordered tug-of-war models suggest that challenges that elicit between-group competition will increase within-group cooperation. Based on social psychology, we note that increased levels of interest and cooperation can be fostered by social norms comparisons in combination with visually compelling representations of individual and collective benchmarks. Finally, we explore how properties of social networks themselves enhance the spread of behaviors through the three degrees rule, homophily, social contagion, and the strength of weak ties. In an age where environmental toxins, habitat loss, population growth, and climate change threaten our future health and survival, we present testable hypotheses and argue for the importance of field experiments to better understand the nexus between the social self, group identity, social networking effects, and potential for supporting collective action via the social Web.

J.L. Dickinson (✉) • R.L. Crain
Citizen Science Program, The Cornell Lab of Ornithology, 159 Sapsucker Woods Road,
Ithaca, NY 14850, USA
e-mail: jld84@cornell.edu

N. Agarwal et al. (eds.), *Online Collective Action*, Lecture Notes in Social Networks,
DOI 10.1007/978-3-7091-1340-0_8, © Springer-Verlag Wien 2014

1   Introduction

The Internet has revolutionized how we learn, invent, and cooperate by decentralizing communications and providing opportunities for individuals to create and display content to an unlimited audience. Given its ability to engender massive collaboration, social learning, and collective intelligence (Boyd and Ellison 2008), the Web presents unprecedented opportunities to understand the human potential for cooperation. The question we raise here is whether, by combining the power of social networking with the sense of purpose found in citizen science, we can provide a new level of support for cooperation within the context of pro-environmental behaviors like energy conservation and habitat restoration. We raise this question at a time when scientific collaborations are occurring over the Internet at previously unprecedented scales, not only in the conservation arena but in a wide variety of disciplines from mathematics with Tim Gowers' Polythmaths blog, where people collaboratively tackle new mathematical proofs (Gowers and Nielsen 2011), to biochemistry, with the game-like citizen science project, Fold.it, whose participants recently discovered a new protein that regulates infection by the Simian AIDS virus (Khatib et al. 2011). The Internet's ability to build large, networked, communities, showcasing the collective impacts of people's contributions and efforts, make the Web the most important tool in history for collaboratively managing public goods.

From an evolutionary standpoint, rapid societal changes in how we cooperate, learn, and invent are fascinating. New models suggest that these changes might go so far as to facilitate more equitable distributions of public goods and avert the tragedy of the commons, even in such difficult arenas as climate change and loss of biodiversity (Bimber et al. 2005). This chapter begins with the premise that science literacy alone does not lead to behavioral change (Osbaldiston and Schott 2011). We explore models and empirical work at the forefront of understanding the social and psychological dimensions of human behavior and place these within the context of the influence of social media. What levels of cooperation are possible in a world made small by Internet-based social networks (Watts and Strogatz 1998), and how might integration of ideological and knowledge networks (e.g., environmentalist and science learning networks) support scientifically informed attitudinal and behavioral shifts? Given that people are influenced not just by friends, but by friends of friends of friends (Christakis and Fowler 2009), how does bringing people within three degrees of separation of a large number of others influence our potential for massive collective action? And can that influence help to support the behavioral change required to address dire environmental problems?

Specifically, we focus on the potential for collective action in citizen science environments that bring both crowd-sourcing approaches and social networking to bear on conservation actions. We consider "first generation" citizen science to be a form of crowd-sourcing with weak institutional governance in which self-selected participants tend to buy into institutionally defined goals and protocols (Wiggins and Crowston 2012). Most current citizen science projects are "first generation" in that they support a joint sense of purpose but do little to connect participants with each other. In contrast, electronic practice networks are "self-organizing, open activity systems focused on shared practice that exists primarily through computer-mediated communication" (Wasko and Faraj 2005). We define "next-generation" citizen science practice networks as having properties of both: (1) channeling a sense of purpose through weak institutional governance, which provides specific goals and protocols for the crowd-sourcing of science-based activities and (2) forming electronic practice networks that allow participants to engage in self-organized cooperative activities via social networking. We propose that when such citizen science practice networks are designed with a collective sense of purpose and with an understanding of barriers to and facilitators of pro-environmental behavior, they will provide new opportunities to manage environmental goods, including common-pool resources, like oil, gas, and water, as well as public goods, like climate, air quality, and biodiversity, which can be shared by everyone.

Progress towards understanding how the Web can be used to support and sustain collective action can only be made by engaging in cross-disciplinary thinking and research. Here we investigate the intersections between the fields of evolutionary biology and social psychology, hoping to stimulate their incorporation into new models and empirical studies of how socially networked citizen science can facilitate collective action. We explore these questions by focusing not on the scientific goals of citizen science but on the ways in which projects can be designed to support scientifically informed environmental behaviors and the scaling-up of individual conservation actions. We pursue this question with the idea that personal shifts in behavior are necessary to our collective future, require social support, and are critical precursors to shifts at higher levels of groups and institutions (Ostrom and Cox 2010).

We bring to this work the basic framework of levels of analysis, seminal to evolutionary studies of social behavior in human and nonhuman animals. We assume that only behaviors that confer inclusive fitness advantages to individuals will be maintained in populations, recognizing that individuals gain by passing on their alleles through both descendant and nondescendant kin (Hamilton 1964). Studies of fitness advantages comprise the ultimate or functional level of analysis, while studies at the proximate level of analysis seek to understand fitness-enhancing mechanisms. While proximate mechanisms may at times appear unselfish, we assume they will only evolve if they positively influence inclusive fitness of individuals. Just as individuals can gain by passing on their alleles through both descendant and nondescendant kin (Hamilton 1964), so can they gain when apparently altruistic acts are repaid by others (Nowak 2006).

In this chapter, we first describe the history of "first generation" citizen science projects and the transition to "next generation" citizen science practice networks, as exemplified by YardMap, which literally puts environmental behaviors on the map. We then highlight the special properties of social mapping environments. Finally, we describe research on collective action and pro-social behavior to derive hypotheses and recommendations for how electronic citizen science practice networks can be designed to support environmental behavior.

2   The Transition from First Generation to "Next Generation" Citizen Science to Support Ecological Research, Conservation, and Management

Citizen science encompasses the many ways in which the public collaborates with professional scientists to conduct a research. Within the context of conservation, first generation citizen science often takes the form of crowd-sourcing to collect large quantities of longitudinal data on the distribution and abundance of organisms across their ranges (Howe 2006). In the USA, birds have played a key role in the development of citizen science methodologies since 1900, when Frank Chapman at the American Museum of Natural History launched the Christmas Bird Count. This and other large-scale citizen science efforts have contributed to scientific understanding of the impacts of human-caused climate and habitat change on birds, and are currently the only way to gain an understanding of the impacts of changes in the distribution and abundance of animals and plants at large and conservation-relevant geographic scales (Dickinson et al. 2010a).

Starting in the 1990s, the Cornell Lab of Ornithology (the Cornell Lab) pioneered the use of Geospatial Web applications to collect citizen science data online, producing dynamic mapping visualizations and developing new computational approaches to analyzing patterns in the data (Sullivan et al. 2009). Today the Cornell Lab engages upwards of 200,000 people who contribute observations to a small, strategic set of projects. These projects set the modern standard for crowd-sourcing of data collection on wild birds as important sentinels for the impacts of environmental change.

Environmental research using citizen science data has uncovered emerging patterns and enabled new public understandings of the leading environmental challenges of the twenty-first century (Dickinson and Bonney 2012). Here we extend this work by asking what "next generation" citizen science might look like and what it might be able to accomplish in the conservation arena. Obvious impacts of connecting participants through social networking include expansion of the public knowledge base and increased potential for social learning. But the most intriguing question is whether socially networked citizen science has a vital role to play in collective actions that could alter the trajectory of "wicked problems" like loss of biodiversity and climate change (Dickinson et al. 2010b; Mankoff et al. 2007; Paulos et al. 2008).

The example we focus on here, called YardMap, is both a citizen science project and a test environment for psychological, sociological, and evolutionary hypotheses for how socially networked citizen science, specifically, and electronic practice networks, generally, can support environmental collective action. YardMap is an online mapping application designed to support and display the activities of individuals within a large, socially networked conservation community. It is focused on managing or restoring habitat for birds and reducing carbon emissions (Fig. 1). While the primary focus in YardMap is on individual (or single family) actions, the cumulative impacts of behaviors like energy use and habitat restoration

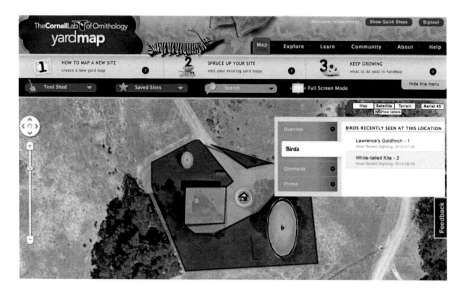

Fig. 1 YardMap application, showing a map with an open site window. Each site and object within a site has an associated information window that feeds comments to and is navigable from the social network

can be significant. For example, Dietz et al. (2009) named 17 simple household actions that, with little in the way of economic or time costs, could save an estimated 123 million metric tons of carbon per year with national, collective implementation. These actions include simple activities like keeping tires properly inflated, cleaning the refrigerator filter once a year, and installing a programmable thermostat. Collective adoption of these practices would represent a decrease of 20 % in household direct emissions or 7.4 % of USA national emissions, a reduction roughly equivalent to the total annual emissions of France.

Individual (or family) scale habitat management for wildlife may also have important cumulative impacts. Daniels and Kirkpatrick (2006) showed that while both landscape-level and garden-level habitat quality predict the presence of Australian bird species in urban environments, small-scale variables, for example, planting natives were stronger predictors of occupancy than the larger, landscape-level variables, like intensity of urbanization. While this pattern suggests that backyard manipulations have a significant impact on Australian bird distributions, the pattern may be the result of spatial clumping of similar gardens, either because people of similar taste and resources tend to live in close proximity or because neighbors engage in imitation and status seeking (Warren 2007). Such spatial autocorrelation would be indicative of potential for strong cumulative impacts of many small, local, and garden-based manipulations. YardMap is designed to observe the behaviors underlying such patterns (e.g., imitation) as well as the patterns themselves, and because YardMap data are integrated with bird-monitoring data, the wildlife outcomes of human behavior and garden-based manipulations can be measured directly.

In 2009, we embarked on creating the YardMap citizen science Web application to support online community involvement in residential conservation practices (Fig. 1); as a "next generation" citizen science platform it is also designed for future tests of hypotheses for which interventions succeed in generating increased levels of cooperation in these two important pro-environmental contexts: energy conservation and habitat restoration.

3 YardMap as an Example of a Citizen Science Practice Network

YardMap is an electronic practice network that allows participants to create, label, and discuss practices by tracing or dragging polygons and objects onto a Google map and commenting on them. The application has a full social network, defined by Boyd and Ellison (2008) as any Web-based service that allows people to (1) create a profile, (2) list others with whom they share a connection, and (3) browse those connections. YardMap allows participants to describe and converse about their mapped conservation practices in a closed social network and to feed news about their practices to Facebook and Twitter to draw others into the project. It also provides educational information relevant to habitat-bird conservation and reduction of carbon emissions with smart tools that point participants to relevant local resources (Fig. 2). Each thing people do in YardMap becomes part of YardMap's database, allowing for longitudinal studies of behavioral change.

Social networking can be built on any form of social connection or interaction, and in the case of YardMap, the connection is rooted in the shared activities of providing information and ideas on personal conservation practices. Activities include sharing self-created maps that display all kinds of choices about use of outdoor spaces and creative, sustainable solutions to common backyard problems (*Where do I put the compost pile? Which natives do well on a step incline?*).

4 The Added Value of Geocollaboration Tools

YardMap is also a form of "geocollaboration" (Hopfer and MacEachren 2007), an area ripe for growth on the Web given its huge potential for creating new pools of knowledge enabling communities to make decisions and organize activity broadly. The emergence of the geospatial Web in the mid-2000s has made projects like YardMap possible (Scharl and Tochtermann 2007). Technologies that enable laypersons to have convenient access to all kinds of georeferenced information are changing the role of maps across the world. Opportunities for "map hacking" (Erle et al. 2005) put control of map-making directly into the hands of the general public. YardMap attempts to access a level of detail about habitat composition and

Fig. 2 YardMap showing cooperative and achievement-based badges displayed in a site window

sustainability practices that typical geographic information systems (GIS) cannot. The types of micro-habitat information collected in YardMap are unprecedented and *depend* on the practices and expertise of individuals, many of which occur at a scale invisible to GIS (Poore 2003), but are still potentially meaningful to birds (Lerman and Warren 2011), insects (Tallamy and Shropshire 2009), and carbon emissions (Dietz et al. 2009).

Maps are particularly amenable to rich, collaborative discussions via social networking. Research has shown that groups have a tendency to make suboptimal decisions because they spend time repeating information already known to all group members rather than utilizing their unique expertise to build a more robust knowledge base from which to make choices (Stasser and Titus 1985). Hopfer and MacEachren (2007) argue that maps help to eliminate this communication bias by *explicitly linking knowledge contributions with geographic objects/locations.*

Map annotations also have the advantage of acting as shared mediating artifacts, sometimes referred to as boundary objects, which orient users, even those with differing cultural knowledge and perspectives, to a common goal. In YardMap, there are many potential layers of map-based artifacts, from an individual tree to a visualization of many YardMappers' efforts to collaboratively create contiguous parcels of viable habitat. Theoretically, homophilous groups will form around these artifacts based on common interest and a desire to collaborate or share information.

Every mark on the map provides an extra-linguistic addition to the collaboration. Text annotations or "notes" and status updates in YardMap provide rich descriptions (e.g., information about intended plans for creating bird habitat or descriptions of how someone managed the transition to more natives in their backyard). Photo-graphs give static detail-rich information (e.g., an image of a tree someone is trying to identify). Drawings, such as habitat polygons and site-lines in YardMap, help to immediately orient collaborators without the use of specialized, extra language. Tags, like information about pesticide use or mowing habits, standardize the vast array of relevant language into accessible 'bits' that become a common lexicon for collaborators coming together for the very first time. These function as straightfor-ward advertisements or "badges" of yard accomplishments (Fig. 2).

Taken together, the entirety of the YardMap's public map comes to represent the collective knowledge base of a community involved in the behavior of creating more sustainable yards and lifestyles. Because of how it is designed, it also reflects the normative behaviors of YardMap participants. As a whole, it is designed to become richer in knowledge resources than the sum of individuals' contributions, likely reflecting both collective intelligence (Woolley et al. 2010) and minable social data (Hill and Terveen 1996). As such, YardMap is an environment that lends itself to exploration and implementation of a wide range of ideas on how online communities can foster collective action.

5 What Game Theoretic Models and Behavioral Games Say About Human Potential for Collective Action

Collective action models have their roots in the zero-contribution thesis of Olson (1965) and in Hardin's (1968) tragedy of the commons, both of which assume that all individuals act as selfish rationalists (rational egoists). Early models included the N-person "Prisoner's Dilemma" game (so named by Dresher 1961) and indicated that rational egoists acting exclusively out of self-interest always do better by defecting rather than cooperating. The Prisoner's Dilemma is a nonzero sum game (one player's gain is not necessarily another's loss), and it is not strictly a cooperative game because players do not interact. The conclusion of these early models was that cooperation will not happen unless individual contributions to a collective action are externally regulated by coercive sanctions, laws, and insti-tutions (Hardin 1968).

Predictions of these early models were not supported by real-life observations nor by a variety of behavioral economics experiments in which participants made decisions about how many "lab dollars" to contribute, knowing that lab dollars represent some fraction of actual dollars they potentially receive at the end of the experiment. Instead of failing to contribute, which would be the optimal decision for a self-interested player, participants tended to contribute between 40 and 60 % of their resources in single-shot Prisoner's Dilemma games (Ostrom 2000). Still, cooperation is limited and in iterated, multiple-trial Prisoner's Dilemmas of finite length, individual contributions decay over time with more than 70 % of partici-pants contributing nothing at all in a salient last round. Even more telling is the observation that people increase their levels of cooperation as they learn and develop a better understanding of how the game works; if they were only self-interested, rational egoists, this would not be so (Ostrom and Ahn 2008).

Many environmental problems are best classified as social dilemmas in which public or fully sharable goods are freely available to noncooperative free-riders. Their solutions require a large crowd to cooperate towards a common goal, and they are characterized by conflict between the public interest and the private interests of individuals. Building on the idea that people are more generous than predicted by early, more restrictive models of cooperation, there has been a gradual movement to introduce a wide range of possibilities for pro-social behavior, beginning with allowing the sorts of interactions that permit people to choose whether and with whom to cooperate (Ostrom 2000). The most important component of second-generation models of collective action is that people are allowed to be "conditional cooperators," eliminating the assumption that everyone is a selfish egoist. The second most important component is that they allow partner choice (Noe 2001).

Pro-social tendencies are evident in observations that people are more likely to cooperate when they believe others will do so, especially when face-to-face communication is added to the mix (Ostrom 2000). Today, many aspects of human pro-social tendencies have been variously accounted for in models of social dilemmas and in behavioral games (the empirical tests of the models). Qualities that variously influence levels of cooperation include cognition (e.g., framing and anchoring; what is most salient) (Critcher and Gilovich 2007; Rand et al. 2009; Tooby et al. 2006) whether or not participants are known to each other; whether they will have opportunities for repeated interactions and reputational display (Barclay 2004; Griskevicius et al. 2010b), leadership effects, visibility of social norms, (Chalub et al. 2006); and whether competition, rewards, or punishment are introduced (Ostrom 2000; Milinski et al. 2006). All of these characteristics can potentially influence cooperative outcomes in electronic social networks.

Models of social dilemmas now typically allow for conditional cooperation in which cooperative tendencies vary among individuals and depend upon whether the actor believes that others will cooperate (Milinski et al. 2002). Unlike first generation models, they are built upon an understanding of checks and balances, including social rewards that appear to govern pro-social behaviors across cultures (Ostrom and Ahn 2008). In a broad sense, second-generation models frame the problem of collective action within the context of social capital, trust networks, and the potential for interpersonal, rather than institutional, rewards and punishment. In a few cases, they even consider *future* benefits of cooperation, such as averting climate change (Santos and Pacheco 2011).

Research on cooperation has yielded general results supporting the idea that, under the right circumstances, cooperation can persist in decentralized communities facilitated by pro-social interactions (Ostrom 2000). In such cases, top down interventions, such as government policies can interfere with, rather than augment, collective action (Montgomery and Bean 1999). On the other hand, we suggest that soft connections to institutions and small "effective group sizes," such as occur with socially networked citizen science, may be very helpful in supporting collective action.

Recently, we have seen a proliferation of evolutionary models that address cooperation and whose outcomes could be influenced by interventions in electronic practice networks. For example, the nested, bordered tug-of-war model places cooperation within the context of intergroup dynamics and has shown that levels of within-group cooperation increase with increasing between-group competition (Reeve and Holldobler 2007). This suggests that interventions that increase between-group competition, in, say, energy conservation, would theoretically increase overall conservation outcomes (Table 1). Another class of models focuses on indirect reciprocity, where altruistic actors receive benefits not directly from recipients but from the observers of their acts (Nowak and Sigmund 1998). This form of cooperation is particularly relevant to electronic practice networks and would be enhanced by providing opportunities for people to display their actions and by providing tools that calculate reputation scores (Table 1) If future benefits are important (Santos and Pacheco 2011), then visualizations that make future gains salient could increase cooperation in electronic practice networks (Table 1). Together, these ideas suggest that building electronic, citizen science practice networks based on new developments in collective action theory, which focus on social rewards and punishment, will extend thinking beyond the simple benefits of the efficiency of social networks (Hampton 2003) to illuminate the Web's potential to support collective action. While considerable attention has been paid to how top-down and institutional structures can support collective action, we are only beginning to investigate how the network structures in decentralized communities matter. Next-generation citizen science clearly provides opportunities for seeding citizen science practice networks with theoretically and empirically informed social drivers of cooperation.

6 How Proximate Mechanisms Can Engender Cooperation in Citizen Science Practice Networks

Before we examine social networking's capacity to augment the potential for collective action, it is essential to understand social and cognitive barriers to pro-environmental behavior and to consider how they can be accounted for in the design of "next generation" citizen science practice networks. Citizen science participants are self-selected. This is important because even when people report having favorable attitudes towards conservation, such as we might expect of people joining pro-environmental citizen science projects, acting on these attitudes and values proves difficult (Kallgren et al. 2000). Increasingly complex cognitive models recognize that people do not make decisions based on information alone, but also weigh social and economic factors (Hines 1987). For example, someone might hear that letting a lawn "go wild" is better for birds, but while she recognizes a connection between a wild lawn and saving songbirds, she is also thinking about her neighbor's negative reaction to her wild lawn and the cost of returning that lawn

Table 1 Testable hypotheses for interventions that will increase pro-environmental behavior in citizen science practice networks

Nature of intervention	Hypotheses for what interventions will increase pro-environmental actions
Provide opportunities to interact socially over map objects	Interacting over map objects and annotations will lead to formation of homophilous groups around those map objects and will orient even those with different cultural knowledges and perspectives towards a common goal
Create opportunities for competition to increase cooperation	Creating opportunities for homiphilous affinity groups within YardMap to compete with each other will increase within-group cooperation, based on the nested tug-of-war model of Reeve and Holldober (2007)
Make future collective gains salient	Creating tools that allow individuals to witness the future collective gains of their actions will increase individual expression of pro-environmental behavior
Self-efficacy hypothesis	Launching challenges with visible and specific benchmarks for individuals will increase expression of pro-environmental behavior
Collective efficacy hypothesis	Launching challenges with visible and specific comparison tools and benchmarks showing the group's collective impacts will increase individual expression of environmental behavior
Reputational display	Providing opportunities for individuals to advertize their pro-environmental actions in the social network through tagging and badges will increase their expression of pro-environmental behaviors
Score-keeping mechanisms	Providing reputational scores for participants using a page-rank-like system or like/dislike and making these scores visible in the social network will increase expression of pro-environmental behaviors
Metrics of allowing	Providing participants with the number of followers they have and making the number of followers visible within the social network serves simultaneously as a visible metric of leadership and gives participants a sense of the reputational consequences of their behaviors; expression of pro-environmental behaviors and leadership (contagion of behaviors) will increase with the number of followers
Visual symbols of following	A picture of eyes next to number of followers will increase individual expression of pro-environmental behaviors; images of eyes have been shown to increase levels of donation to a communal coffee till in comparison with control pictures of flowers (Bateson et al. 2006)
Combine social norms comparisons with benchmarks	Visual images that show where people are relative to social norms, and that simultaneously show benchmarks or goals that are more ambitious than the social norm, will increase expression of pro-environmental behavior and shift the social norm in a more favorable direction (see YardMap pyramid in Fig. 3)

to its tamed state when she has to sell the house (Kurz and Baudains 2011). As human beings respond to social, not just material consequences of their actions, social support for conservation action is likely to be important. This suggests that the current design of YardMap, which allows individuals to "like" each other's actions, will increase enactment of "green" behaviors.

Self-efficacy, which correlates closely with identity and self-esteem (Schwartz et al. 2005), is widely recognized as an important driver of behavioral change. Research indicates that people in the USA believe they have little self-efficacy when it comes to influencing the environment (Gupta and Ogden 2009). Such beliefs lead people to assume that responsibility for environmental conservation must fall to larger entities, such as government or business (Wray-Lake et al. 2010). Activities focused on problems, rather than solutions, may exacerbate this. For example, calculating a carbon footprint, as is popular on myriad websites these days, may reinforce feelings of insignificance. We suggest that launching challenges with specific benchmarks and providing mechanisms for people to track their progress relative to explicit goals, will increase sense of self-efficacy in YardMap (Table 1). We also hypothesize that emphasizing group efficacy will enhance feelings of self-efficacy and support pro-environmental behavior in YardMap. For example, YardMap attempts to mitigate a lack of self-efficacy by providing scientifically informed statistics and information on the collective impacts of specific actions and by enabling participants to invite others to join in. As increasing numbers of people begin to adopt beneficial practices, the collective impacts can be displayed in compelling visualizations (Table 1).

In Web social networks, projects aimed at generating change must consider the potential to shift social norms, which profoundly influence what people are likely to do. Altruism itself is a social norm; thus, campaigns that foreground messages about environmentalism as a shared, social value should find some success in motivating behavioral change (Goldstein et al. 2008). The norm of reciprocity also appears to maintain social relationships across cultures, and thus should be a powerful motivator in pro-environmental contexts (Cialdini and Goldstein 2004). These and other ideas have been the focus of a series of experiments placing signs in hotel rooms to see which sorts of messages will get people to participate in hotel towel reuse. When the hotel made a donation to an environmental organization before inviting guests to reuse their towels, reuse rose to 42.5 %, compared to 30 % for appeals based more generally on cooperation, social responsibility, or the environment (Goldstein et al. 2007). This suggests that norms of reciprocity are somewhat effective at increasing rates of hotel towel reuse. In a later study, appeals highlighting descriptive norms were similarly successful (44 % reuse), in this case, letting people know that 75 % of the hotel's guests reused their towels. The more substantial increase to 49.3 % was not seen until researchers invoked commonality by producing signs saying that 75 % of people staying in "this hotel room" have reused their towels (Burger et al. 2004; Goldstein et al. 2008). In YardMap, similar effects may be achieved using smart, spatially aware tools that show sustainability achievements of nearby others.

While the hotel towel study demonstrates the power of social norms at eliciting cooperation, another study involving energy consumption illustrates a key drawback of social norms comparisons. While enabling customers to see their energy use in comparison to their neighbors' use altered consumption, high-use households showed strong movement downwards towards the norm, while low use households showed a tendency to increase their energy use when they could see that they were using less than the average amount. This demonstrates that simple quantitative comparisons can have a leveling effect, keeping the population in the vicinity of the existing social norm, rather than shifting both the population and the social norm in the desired direction. We suggest that this leveling effect may be ameliorated in YardMap by using visualizations that combine social norms comparisons with directional benchmarks (Table 1, Fig. 3) or by combining images of social approval (e.g., a smiling face) with visual representations of the most pro-environmental benchmarks.

As Rosenberg (2011) notes, people do not change their behaviors based on new information or fear appeals but by encountering a new peer group with which they can identify and within which social norms and new behaviors are seen as "really cool" and even heroic. Given that people are highly influenced by comparison with others (Frank 2011), a key question as yet unresolved is how social comparison can facilitate shifting of social norms and whether social norms shift more rapidly in electronic practice networks than in offline communities. In YardMap, campaigns that make competition and social comparison of "greenness" salient may drive increased investment in pro-environmental behaviors with challenges, such as, "How green can you be?"

7 Designing Citizen Science Practice Networks to Support Pro-Environmental Behavior

The design of citizen science practice networks to support pro-environmental behavior is an area ripe for empirical research. We propose that pro-social tendencies necessary to support collective action are activated once participants are (1) linked to one another within an online social network, (2) can witness what others are doing, and (3) can display their actions socially through e-friendships. These provide opportunities for social identity and normative influence, as well as reputation maintenance, formation of trust networks, and delivery of social rewards and punishment required to prevent free-riding (Table 1). We also suggest that properties of social networks, themselves, will augment the potential for collective action at meaningful scales.

On online social networks, spread of behavior can be influenced by the sheer number of connections people have, the visibility of their connections, and the ability to form very specific connections based on similarity (homophily) or status (e.g., centrality in the network) (Ostrom and Ahn 2008). While electronic practice

Fig. 3 Dynamic tool
showing YardMap
participant where he or she
fits relative to low, average,
above average, and top
participants

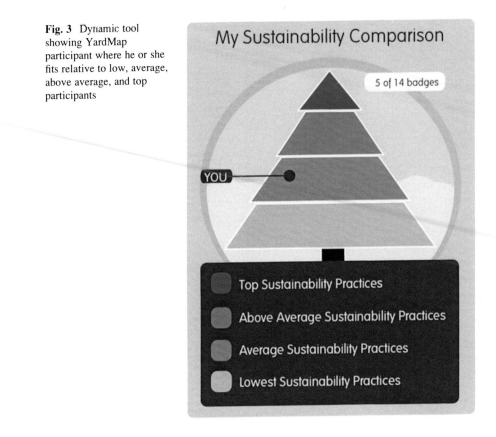

networks like YardMap will likely be quite homophilous, it remains to be seen whether they can achieve the critical mass to make a difference. When people have a large number of connections, they tend to have high exposure, and so they become influential; in addition to this, some individuals are behavioral opinion leaders, while others have outsized influence because they are connected to a few key leaders whom they can influence (Easley and Kleinberg 2010, p. 513). On the other hand, individuals also tend to be most strongly connected to similar others and they will tend to share a large proportion of overlapping friendships with such people (McPherson et al. 2001). This fosters the maintenance of cooperation in social networks simply because people are more likely to connect with and imitate others like them (Durrett and Levin 2005). Research in social psychology has shown that compliance is more likely to occur among similar people, even when the shared trait is as simple as a first name, a fingerprint, or birthday (Burger et al. 2004). It is possible to distinguish influence-based contagion from homphily-driven diffusion, and this becomes important in determining which design strategies will counter versus facilitate the spread of behaviors (Aral et al. 2009). A hypothesis in need of testing is whether adding tools that allow participants to form homophilous subgroups will increase diffusion of cooperative behaviors in citizen science practice networks like YardMap or whether the entire network is so homophilous that its structure has little influence (Table 1).

While strong ties among similar individuals are certainly important in creating "bonding" social capital, people interacting in large, electronic social networks also tend to have weak ties to a large number of less familiar individuals. Such bridging social capital helps to generate rapid diffusion of information and behaviors throughout the larger network (Backstrom et al. 2006), while also tapping into diverse skills and opinions, which likely fosters the emergence of collective intelligence and innovation (Woolley et al. 2010). Given that YardMap taps into gardening, a popular hobby that crosses socioeconomic and cultural barriers, there may be more potential for bridging social capital than would exist in social networks focused only on sustainability.

In the end, network density may be even more critical. The density of networks on the whole is thought to increase the potential for collective action (Marwell et al. 1988). This is corroborated by recent simulations suggesting that high levels of cooperation can be achieved, even without reputation maintenance, if the benefit: cost ratio of cooperation exceeds the average connectivity (Ohtsuki et al. 2006).

The transparency of electronic practice networks can enable pro-social mechanisms known to generate and maintain collective action, including group identity, reputation, social capital, and normative behavior. Various models of collective action suggest that reputation and opportunities to detect cheaters promote cooperation (Hauert et al. 2002). Additionally, we know from social psychology research that activities that give people a sense of purpose and belonging, or opportunities for identity-building and reputation maintenance, also facilitate cooperation (Griskevicius et al. 2010a). Understanding the nature of online reputations and the potential for deception is thus critical to assessing the potential for collective action in electronic practice networks.

Reputation may be established and communicated broadly within social networks, not only through direct connections (friends) but also along a variety of paths to friends of friends, but success of online collective action will depend on the accuracy of reputational information and the trustworthiness of individuals. In face-to-face social networks, the accuracy of information about people degrades as it becomes second-hand information, often leading to more extreme impressions and, potentially, judgments (Gilovich 1987). In contrast, in online social networks, information about an individual is usually replicated precisely as it is conveyed from friends to friends of friends through sharing tools. This should allow reputations to be transmitted more accurately than in "real life."

Most modern models of cooperation rely on trust and indicate that the potential for cooperation is enhanced by reputational display and scorekeeping tools (Nowak and Sigmund 1998). Early on, social networking tools provided opportunities for "referral trust" in which individuals could recommend others as trustworthy (Artz and Gil 2007). But contexts for trust differ, for example, a participant in YardMap could score high in terms of interpersonal trustworthiness and low in terms of the accuracy of plant identifications he or she conveys. Much of the emphasis on trust has been on trustworthiness of information, but a variety of tools, including computational metrics, have been developed to combine different aspects of reputation into a single score. These include tools that rank interpersonal trustworthiness based on people's position in the social network, for example, their degree or their degree weighted by the trustworthiness of their connections as sorted and assigned by the PageRank algorithm (Hogg and Adamic 2004). Refinement of such methods will continue to improve precision and accuracy of various metrics of trust and trustworthiness in social networks (Table 1).

Game theoretic models have shown that accessible information on reputation makes social interactions more efficient; this efficiency could be amplified in social networks, especially given the high speed of transactions (Raub and Weesie 1990). Mechanisms that allow participants to assess trustworthiness and that do not allow cheating seem feasible in online social networks and will likely be required if electronic practice networks like YardMap are to foster collective action (Table 1). Social networking tools can also summarize information about the behavior of others via computational algorithms (e.g., scores of trustworthiness, Hogg and Adamic 2004). When actions are reported and displayed in a highly visual Web interface, they are potentially searchable and visible to all, which can aid collective action (Table 1). Additionally, the ability to pay rewards and mete out punishment offers significant, unrealized potential to foster collective action; evidence suggests that in online interactions involving markets (Bolton et al. 2004), negative feedback has more impact than positive feedback, but this idea has recently been contested within the context of collective action (Rand and Nowak 2011). Despite evidence that deception occurs in social networks just as in real life (Toma and Hancock 2012), we argue that the social Web has sufficient corrective potential to enable the proximate mechanisms required to generate massive collective action.

8 Shifting Attitudes, Behaviors, and Social Norms within Electronic Practice Networks

Despite extensive, expensive efforts aimed at getting people to adopt pro-environmental practices, we have seen few major collective shifts in behavior (McKenzie-Mohr and Oskamp 1995). This is true despite the massive proliferation and public adoption of Web 2.0 technologies in the last 10 years. It is not for lack of trying—there are *countless* websites devoted to "going green." We suggest that one problem has been a lack of integration of research from evolutionary biology and social psychology to inform the design and development of such Web interventions. Exploring the interaction of social networking with proximate mechanisms governing collective action can help to test and guide design principles for citizen science practice networks.

Here we illustrate such design principles using the new citizen science project, YardMap.org, which entered beta testing with the public in 2012. We propose that the potential for cooperation is enhanced by combining social network features with features that enable the full range of behavioral mechanisms that support cooperation, based on previous models and experiments (Fig. 2, Table 1). These mechanisms not only include reputational drivers of cooperation but also acknowledge that between-group competition can lead to increased within-group cooperation (Reeve and Holldobler 2007). Projects like YardMap can be used to conduct "field experiments" within electronic practice networks to measure which mechanisms are well supported and to examine the outcomes of cooperation through electronic tracking of conservation behaviors, social network structure, and the form and nature of interactions within the networks.

In Table 1, cited throughout, we highlight recommendations for further research to test hypotheses for which interventions or design features will support cooperation in pro-environmental contexts. Citizen science practice communities like YardMap, when informed by evolutionary and social psychology, serve as real life labs for testing interventions that are thought to promote collective action while simultaneously testing the impacts of scientifically informed pro-environmental practices. Our synthesis of research suggests that projects will do best to employ tactics that promote normative visibility in combination with benchmarks (Fig. 3), emphasize to users that their actions are both socially and practically valuable, and show how the cumulative impacts of participants' actions can be significant (Table 1).

The question of how the Internet might help in the effort to manage collective actions is an important one, especially within the context of the dire environmental problems of the twenty-first century. Evidence suggests that the most fruitful course will be to structure online environments to allow for expression of the full range of proximate pro-social mechanisms that have promoted cooperation and altruism throughout human evolution.

Acknowledgements We thank Walter Koenig and two anonymous reviewers for comments on this chapter. Our work is supported by NSF (DRL) ISE grant # 0917487.

References

Aral S, Muchnik L, Sundararajan A (2009) Distinguishing influence-based contagion from homphily-driven diffusion in dynamic networks. Proc Natl Acad Sci U S A 106:21544–21549

Artz D, Gil Y (2007) A survey of trust in computer science and the semantic Web. Web Semant Sci Serv Agents World Wide Web 5:58–71

Backstrom L, Huttenlocher D, Kleinberg J, Lan X (2006) Group formation in large social networks: membership, growth, and evolution. In: Proceedings of 12th ACM SIGKDD international conference on knowledge discovery and data mining. Philadelphia, PA, 20–23 Aug 2006

Barclay P (2004) Trustworthiness and competitive altruism can also solve the "tragedy of the commons". Evol Hum Behav 25:209–220

Bateson M, Nettle D, Roberts G (2006) Cues of being watched enhance cooperation in a real-world setting. Biol Lett 2:412–414

Bimber B, Flanagin AJ, Stohl C (2005) Reconceptualizing collective action in the contemporary media environment. Commun Theory 15:365–388

Bolton GE, Katock E, Ockenfels A (2004) How effective are electronic reputation mechanisms? An experimental investigation. Manag Sci 50:1587–1602

Boyd D, Ellison N (2008) Social network sites: definition, history, and scholarship 13:210–230. doi: 10.1111/j.1083-6101.2007.00393.x

Burger JM, Messian N, Patel S, del Prado A, Anderson C (2004) What a coincidence! The effects of incidental similarity on compliance. Personal Soc Psychol Bull 30:35–43

Chalub FACC, Santos FC, Pacheco JM (2006) The evolution of norms. J Theor Biol 241:233–240

Christakis NA, Fowler JH (2009) Connected: the surprising power of our social networks and how they shape our lives, Little, Brown & Company, New York

Cialdini RB, Goldstein NJ (2004) Social influence: compliance and conformity. Annu Rev Psychol 55:591–621

Critcher CR, Gilovich T (2007) Incidental environmental anchors. J Behav Decis Mak 21:241–251

Daniels GD, Kirkpatrick JB (2006) Does variation in garden characteristics influence the conservation of birds in suburbia? Biol Conserv 133:326–335

Dickinson JL, Bonney R (2012) Citizen science: public collaboration in environmental research. Cornell University Press, Ithaca, NY

Dickinson JL, Zuckerberg BZ, Bonter DN (2010a) Citizen science as an ecological research tool: challenges and benefits. Annu Rev Ecol Evol Syst 41:149–172

Dickinson JL, Zuckerberg BZ, Bonter DN (2010b) Citizen science as an ecological research tool: challenges and benefits. Annu Rev Ecol Evol Syst 41:149–172

Dietz T, Gardner G, Gilligan J, Stern P, Vandenbergh M (2009) Household actions can provide a behavioral wedge to rapidly reduce US carbon emissions. Proc Natl Acad Sci U S A 106(44): 18452

Dresher M (1961) The mathematics of games of strategy: theory and applications. Prentice-Hall, Englewood Cliffs, NJ

Durrett R, Levin SA (2005) Can stable social groups be maintained by homophious imitation alone? J Econ Behav Organ 57:267–286

Easley D, Kleinberg J (2010) Networks, crowds, and markets. Cambridge University Press, Cambridge, England

Erle S, Gibson R, Walsh J, Laurent SS (eds) (2005) Mapping hacks: tips & tools for electronic cartography. O'Reilly, Aptos

Frank R (2011) The Darwin economy: liberty, competition, and the common good. Princeton University Press, Princeton, NJ

Gilovich T (1987) Secondhand information and social judgment. J Exp Soc Psychol 23:59–74

Goldstein NJ, Griskevicius V, Cialdini RB (2007) Invoking social norms: a social psychology perspective on improving hotels' linen-reuse programs. Cornell Hotel Restaur Admin Q 48: 145–150

Goldstein NJ, Cialdini RB, Griskevicius V (2008) A room with a viewpoint: using social norms to motivate environmental conservation in hotels. J Consum Res 35:472–482. doi:10.1086/586910

Gowers T, Nielsen M (2011) Massively collaborative mathematics. Nature 461:879–881

Griskevicius V, Tybur JM, Van den Bergh B (2010a) Going green to be seen: status, reputation, and conspicuous conservation. J Pers Soc Psychol 98(3):392–404. doi:10.1037/a0017346

Griskevicius V, Tybur JM, van den Bergh B (2010b) Going green to be seen: status, reputation, and conspicuous conservation. J Pers Soc Psychol 98:392–404

Gupta S, Ogden DT (2009) To buy or not to buy? A social dilemma perspective on green buying. J Consum Mark 26(6):376–391. doi:10.1108/07363760910988201

Hamilton WD (1964) The genetical evolution of social behavior. I and II. J Theor Biol 7:1–52

Hampton KN (2003) Grieving for a lost network: collective action in a wired suburb. Inf Soc 19:417–428

Hardin G (1968) The tragedy of the commons. Science 162:1243–1248

Hauert C, De Monte S, Hofbauer J, Sigmund K (2002) Volunteering as red queen mechanism for cooperation in public goods games. Science 296:1129–1131

Hill WC (1996) Terveen LG using frequency-of-mention in public conversations for social filtering. In: Proceedings of CSCW'96. ACM, New York, pp 106–112

Hines JM (1987) Analysis and synthesis of research on responsible environmental behavior: a meta-analysis. J Environ Educ 18:1–8

Hogg T, Adamic L (2004) Enhancing reputation mechanisms via online social networks. Paper presented at the ACM conference on electronic commerce. New York, NY, 17–20 May, 2004

Hopfer S, MacEachren AM (2007) Leveraging the potential of geospatial annotations for collaboration: a communication theory perspective. Int J Geogr Inf Sci 21(8):921–934. doi:10.1080/13658810701377780

Howe G (2006) The rise of crowd-sourcing. Wired.

Kallgren CA, Reno RR, Cialdini RB (2000) A focus theory of normative conduct: when norms do and do not affect behavior. Personal Soc Psychol Bull 26(8):1002–1012. doi:10.1177/01461672002610009

Khatib F, DiMaio F, Group FC, Group FVC, Cooper S, Kazmierczyk M, Gilski M, Krzywda S, Zabranska H, Pichova I, Thompson J, Popović Z, Jaskolski M, Baker D (2011) Crystal structure of a monomeric retroviral protease solved by protein folding game players. Nat Struct Mol Biol 18:1175–1177

Kurz T, Baudains C (2011) Biodiversity in the front yard: an investigation of landscape preference in a domestic urban context. Environ Behav Online, 7 Nov, 2010. doi:10.1177/0013916510385542

Lerman SB, Warren PS (2011) The conservation value of residential yards: linking birds and people. Ecol Appl 21:1327–1339

Mankoff J, Matthews D, Fussell SR, Johnson M (2007) Leveraging social networks to motivate individuals to reduce their ecological footprints. In: 40th Hawaii international conference on systems science. IEEE Computer Society, Waikoloa, Big Island, Hawaii, 3–6 Jan 2007

Marwell G, Oliver PE, Prahl R (1988) Social networks and collective action: a theory of the critical mass. III. Am J Sociol 94:502–534

McKenzie-Mohr D, Oskamp S (1995) Psychology and sustainability: an introduction. J Soc Issues 51(4):1–14. doi:10.1111/j.1540-4560.1995.tb01345.x

McPherson M, Smith-Lovin L, Cook JM (2001) Birds of a feather: homophily in social networks. Annu Rev Sociol 27:415–444

Milinski M, Semmann D, Krambeck H (2002) Reputation helps solve the 'tragedy of the commons'. Nature 415:424–426

Milinski M, Semmann D, Krambeck H-J, Marotzke J (2006) Stabilizing the earth's climate is not a losing game: supporting evidence from public goods experiments. Proc Natl Acad Sci U S A 103:3994–3998

Montgomery MR, Bean R (1999) Market failure, government failure, and the private supply of public goods: the case of climate-controlled walkway networks. Public Choice 99:403–437

Noe R (2001) Biological markets: partner choice as the driving force behind the evolution of mutualisms. In: Noe R, van Hoof JARAM, Hammerstein P (eds) Economics in nature: social dilemmas, mate choice, and biological markets. Cambridge University Press, Cambridge, pp 93–118

Nowak MA, Sigmund K (1998) Evolution of indirect reciprocity by image scoring. Nature 393:573–577

Nowak MA (2006) Five rules for the evolution of cooperation. Science 314:1560–1563

Ohtsuki H, Hauert C, Lieberman E, Nowak MA (2006) A simple rule for the evolution of cooperation in graphs and social networks. Nature 441:502–505

Olson M (1965) The logic of collective action: public goods and the theory of groups. Harvard University Press, Cambridge, MA

Osbaldiston R, Schott JP (2011) Environmental sustainability and behavioral science: meta-analysis of proenvironmental behavior experiments. Environ and Behav [Online ahead of print]. doi:10.1177/0013916511402673

Ostrom E (2000) Collective action and the evolution of social norms. J Econ Perspect 14:137–158

Ostrom E, Ahn TK (2008) The meaning of social capital and its link to collective action. In: Svendsen GT, Svendsen GL (eds) Handbook of social capital: the troika of sociology, political science and economics. Edward Elgar, Northampton, MA, pp 17–35

Ostrom E, Cox M (2010) Moving beyond panaceas: a multi-tiered diagnostic approach for social-ecological analysis. Environ Conserv 37:451–463

Paulos E, Honicky RJ, Hooker B (2008) Citizen science: enabling participatory urbanism. In: Foth M (ed) Handbook of research on urban informatics: the practice and promise of the real-time city. IGI Global, Hershey, PA

Poore BS (2003) The open black box: the role of the end-user in GIS integration. Can Geogr 47(1): 62–74. doi:10.1111/1541-0064.02e13

Rand G, Nowak MA (2011) The evolution of antisocial punishment in public goods games. Nat Commun 2:434

Rand G, Dreber A, Ellingsen T, Fudenberg D, Nowak MA (2009) Positive interactions promote public cooperation. Science 325:1272–1275

Raub W, Weesie J (1990) Reputation and efficiency in social interactions: an example of network effects. Am J Sociol 96:626–654

Reeve HK, Holldobler B (2007) The emergence of a superorganism through intergroup competition. Proc Natl Acad Sci U S A 104:9736–9740

Rosenberg T (2011) Join the club: how peer pressure can transform the world. W. W. Norton & Company, New York

Santos FC, Pacheco JM (2011) Risk of collective failure provides an escape from the tragedy of the commons. Proc Natl Acad Sci U S A 108:10421–10425

Scharl A, Tochtermann K (2007) The geospatial web: how geobrowsers, social software and the web 2.0 are shaping the network society. Springer, London

Schwartz SJ, Cote JE, Arnett JJ (2005) Identity and agency in emerging adulthood: two developmental routes in the individualization process. Youth Soc 37:201–209

Stasser G, Titus W (1985) Pooling of unshared information in group decision making: biased information sampling during discussion. J Pers Soc Psychol 48(6):1467–1478

Sullivan BL, Wood CL, Iliff MJ, Bonney RE, Fink D, Kelling S (2009) eBird: a citizen-based bird observation network in the biological sciences. Biol Conserv 142(10):2282–2292. doi:10.1016/j.biocon.2009.05.006

Tallamy DW, Shropshire K (2009) Ranking lepidopteran use of native versus introduced plants. Conserv Biol 23(4):941–947. doi:10.1111/j.1523-1739.2009.01202.x

Toma C, Hancock JT (2012) What lies beneath: the linguistic traces of deception in online dating profiles. J Commun 62:78–97

Tooby J, Cosmides L, Price ME (2006) Cognitive adaptations for n-person exchange: the evolutionary roots of organizational behavior. Manag Decis Econ 27:103–129

Warren P (2007) Plants of a feather: spatial autocorrelation of gardening in suburban neighborhoods. Biol Conserv 141:3–4

Wasko MM, Faraj S (2005) Why should I share? Examining social capital and knowedge contribution in electronic networks of practice. MIS Q 29:35–57

Watts DJ, Strogatz SH (1998) Collective dynamics of 'small-world' networks. Nature 393:409–410

Wiggins A, Crowston K (2012) Developing a conceptual model of virtual organizations for citizen science. In: Agrawal N, Lim M, Wigand R (eds) Online collective action: dynamics of the crowd in social media. Springer, New York

Woolley AW, Chabris CF, Pentland A, Hashmi N, Malone TW (2010) Evidence for a collective intelligence factor in the performance of human groups. Science 330:686–688

Wray-Lake L, Flanagan CA, Osgood DW (2010) Examining trends in adolescent environmental attitudes, beliefs, and behaviors across three decades. Environ Behav 42(1):61–85. doi:10.1177/0013916509335163

Part III
Case Studies

The Spanish "Indignados" Movement: Time Dynamics, Geographical Distribution, and Recruitment Mechanisms

Javier Borge-Holthoefer, Sandra González-Bailón, Alejandro Rivero, and Yamir Moreno

Abstract Online social networks have an enormous impact on opinions and cultural trends. Also, these platforms have been revealed as a fundamental organizing mechanism in country-wide social movements. Recent events in the Middle East and North Africa (the wave of protests in the Arab world), across Europe (in the form of anti-cuts demonstrations or riots) and in the United States have generated much discussion on how digital media is connected to the diffusion of protests. In this chapter, we investigate, from a complex network perspective, the mechanisms driving the emergence, development and stabilization of the "Indignados" movement in Spain, analyzing data from the period between April 25 and May 26, 2011. Using 70 keywords related to the movement, we analyze 581,749 Twitter messages coming from 87,569 users. The online trace of the 15M protests provides a unique opportunity to tackle central issues in the social network

J. Borge-Holthoefer (✉)
Instituto de Biocomputación y Física de Sistemas Complejos (BIFI), Universidad de Zaragoza, Zaragoza, Spain

Qatar Computing Research Institute, 13th Floor, Tornado Tower, P.O. Box 5825, Doha, Qatar
e-mail: borge.holthoefer@gmail.com; jborge@qf.org.qa

S. González-Bailón
Annenberg School for Communication, University of Pennsylvania, Philadelphia, PA, USA
e-mail: sgonzalezbailon@asc.upenn.edu

A. Rivero
Instituto de Biocomputación y Física de Sistemas Complejos (BIFI), Universidad de Zaragoza, Zaragoza, Spain
e-mail: arivero@bifi.es

Y. Moreno
Instituto de Biocomputación y Física de Sistemas Complejos (BIFI), Universidad de Zaragoza, Zaragoza, Spain

Department of Theoretical Physics, Faculty of Sciences, Universidad de Zaragoza, Zaragoza 50009, Spain
e-mail: yamir.moreno@gmail.com

N. Agarwal et al. (eds.), *Online Collective Action*, Lecture Notes in Social Networks, DOI 10.1007/978-3-7091-1340-0_9, © Springer-Verlag Wien 2014

literature like recruitment patterns or information cascades. These findings shed light on the connection between online networks and social movements and offer an empirical test to elusive sociological questions about collective action.

1 Introduction

Modern online socio-technological systems are changing the way we communicate with each other, revolutionizing the methods researchers have at hand to face old sociological questions. The networked nature of online communication platforms, plus the inherent complexity of the activity within them, places modern network theory (Boccaletti et al. 2006) as an outstanding tool to provide deeper insight into questions such as the emergence of authoritative or privileged nodes (Gonçalves et al. 2011; Ratkiewicz et al. 2010), the importance of the strength and range of connections (Onnela et al. 2007) or time patterns of human activity (Barabási 2005). These platforms generate an enormous amount of time-stamped data, making it possible for the first time to study the fast dynamics associated with different spreading processes at a global scale. These novel and rich data sources allow testing different social dynamics and models that would otherwise be highly elusive with traditional data-gathering methods. Additionally, the availability of data enables the study of phenomena that take place on timescales ranging from a few minutes or hours to years.

This chapter focuses on how social networking sites (SNSs) contribute to the organization of collective action. Online social media provide efficient and fast means to group together many actors around a common cause, as exemplified by the wave of social unrest staged around the world in 2011. Protests have arisen in the aftermath of financial and political crises in the form of pacific civil movements (as with the Spanish "Indignados" in May or "Occupy Wall Street" in the United States, which culminated in global marches on October 15); economy-related demonstrations with some violent episodes (as in the case of the protests in Greece, Italy, and UK); and regime-changing (and often violent) political uprisings (as in the "Arab spring"). Although these protests respond to very different circumstances, they all made use of SNSs to help protesters self-organize.

We focus on one of these examples: the Spanish "Indignados" movement (also known as the "15M" movement), which was still active as of January 2012. After the original mass demonstrations on May 15 (day D from now on), hundreds of participants decided to continue the protests camping in the main squares of several cities (Puerta del Sol in Madrid, Plaça de Catalunya in Barcelona) until May 22, the following Sunday and the date for regional and local elections. During that week, protesters created committees to coordinate the logistics of campsites and organized around open popular assemblies. The media, which had not covered the movement until the day of the first big demonstrations, started covering the protests on a daily basis, particularly after the authorities tried to evict protesters from the squares by force, and the Electoral Committee declared the protests illegal. Despite the prohibition, the camps remained in place, receiving increasing popular support and staging daily demonstrations. As many of the adherents were online social media users, the growth and stabilization of the movement can be analyzed using time-stamped data of Twitter messages, which we have collected and analyzed as described below (Sects. 3 and 4).

A social phenomenon like the 15M movement is an excellent opportunity to understand network formation processes and online spreading dynamics. We characterize the structural patterns of the network of users who sent or received tweets containing keywords related to the 15M movement. We find that this network displays the typical features of other networks in nature such as scale-free degree distributions, a community structure at the mesoscale and high structural robustness (Boccaletti et al. 2006; Dorogovtsev et al. 2008). Beyond this general characterization—which has deep implications regarding the dynamical processes occurring on top of the structure—we also analyze two key sociological aspects: First, we attempt to understand how recruitment takes place in a socio-political movement such as 15M, providing novel empirical evidence of the mechanisms behind the decision to join a protest and enroll in collective action. Models of collective action have identified important network mechanisms behind the decision to join a protest, but they suffer from lack of empirical calibration and external validity. Online networks, and the role that SNSs play in articulating the growth of protests, offer a great opportunity to explore recruitment mechanisms in an empirical setting. Second, we analyze the dynamical patterns that characterize the spreading of information over the 15M network. We show that diffusion is hampered by what we called *information sinks*: A great part of the traffic is delivered to a few users that do not pass the information along. We also find that because of the firewalls created by these *information sinks*, very few messages generate global cascades. This falls in line with related research, which has shown that information cascades in online networks occur only rarely (Bakshy et al. 2011; Kwak et al. 2010; Sun et al. 2009), with the implication that even online it is difficult to reach and mobilize a high number of people.

Details of these analyses follow, capitalizing from and synthesizing previous work (Borge-Holthoefer et al. 2011, 2012; Gonzalez-Bailón et al. 2011) where partial stances of the phenomenon were offered. Prior to that, the following section presents a succinct summary of the principles of modern network theory.

2 Network Theory in a Nutshell

Historically, the study of networks has been mainly the domain of a branch of discrete mathematics known as graph theory. Since its birth in 1736, when the Swiss mathematician Leonhard Euler published the solution to the Königsberg bridge problem (consisting in finding a round trip that traversed each of the bridges of the Prussian city of Königsberg exactly once), graph theory has witnessed many exciting developments and has provided answers to a series of practical questions. In addition to the developments in mathematical graph theory, the study of networks has seen important achievements in some specialized contexts, as for instance in the social sciences. Social network analysis started to develop in the early 1920s and focuses on relationships among social entities as communication between members of a group, trades among nations or economic transactions between corporations. It was in the second half of the twentieth century, however, when networks began to produce fruitful models, as for instance in the works by Rapoport (1953, and successive) or by Granovetter (1973), and some methodological advances were made (Freeman 1977).

Although the concept of small world was already well known by sociologists (Milgram 1967; Travers and Milgram 1969), it was in 1998 that Watts and Strogatz formalized the properties of small-world networks (Watts and Strogatz 1998), a model that became the seed for the modern theory of complex networks. The final leap to this emerging subfield was encouraged by increasing data availability and computing capacity, which were made widely available in the 1990s. From then on, network analysis methods have been used to model many complex real-world phenomena. Examples are numerous, ranging from the Internet—a network of routers or domains—to the global economy—a network of national economies, which are themselves networks of markets—or energy, which is distributed through transportation networks, both in living organisms, man-made infrastructures and in many physical systems such as the power grids. All of them stand as examples of research under a network modelling approach.

There are excellent reviews devoted to the structural characterization and evolution of complex networks (Albert and Barabási 2002; Boccaletti et al. 2006; Dorogovtsev et al. 2008; Newman 2003). Here, we introduce only those network descriptors that are mentioned along the chapter (see, e.g., Table 1).

The mathematical abstraction of a complex network is a graph G comprising a set of N nodes (or vertices) connected by a set of E links (or edges), k_i being the degree (number of links) of node i. This graph is represented by the adjacency matrix A, with entries $a_{ij} = 1$ if a directed link from i to j exists and 0 otherwise. In the more general case of a weighted network, the graph is characterized by a matrix W with entries w_{ij} representing the strength (or weight) of the link from i to j. A connected network is an undirected network such that there exists a path between all pairs of vertices. If the network is directed and there exists a path from each vertex to every other vertex, then it is a strongly connected network. If the network is not connected, i.e. it is made up of broken fragments, an interesting question is how large the giant component (or the largest connected subgraph) is with respect to N.

Much of the work in network theory deals with cumulative degree distributions $P(k)$. A plot of $P(k)$ for any given network is built through a cumulative histogram of the degrees of vertices, and this is the type of plot used throughout this article (and often referred to as just "degree distribution"). Although the degree of a vertex is a local quantity, a cumulative degree distribution often determines some important global characteristics of networks. This notion can be extended to the weighted scheme, see Fig. 3. The first classification of complex networks is related to the

Table 1 Topological descriptors for both the static and dynamic view of the Twitter data

		Static network	Activity network
N	Number of nodes	87,569	87,569
E	Number of edges	6,030,459	206,592
$<k>$	Average degree	69	2.36
C	Clustering coefficient	0.22	0.034
L	Average path length	3.24	1.7
D	Diameter	11	4
r	Assortativity	-0.14	0.005
SCC	Number of strongly connected components	5,249	73,389
$Nscgc$	Size of giant component	82,253	13,103
$max(k_{in})$	Maximum in-degree	5,773	29,155
$max(k_{out})$	Maximum out-degree	31,798	289

The information in this table for the activity network corresponds to the accumulation of directed (@) messages up to $D+10$. Both networks are built from the same users, but descriptors diverge largely. Remarkably, although both networks are sparse (low $<k>$ with respect to the system's size N), dynamic activity is much more so. Many nodes, however, act as information sinks—they never emit messages, although most messages are mentions to them (Borge-Holthoefer et al. 2011). The clustering coefficient C is significantly high in both networks, given their density, which, in combination with the low L and D suggests that the network has a "small world" structure. Unlike other reported cases (Newman 2003), r is negative for the static network, whereas degree appears to be uncorrelated in the case of the activity network. The giant strongly connected component $Nscgc$ is comparable to the system's size in the follower network, which means that almost every node in it is mutually reachable; that is not the case for the dynamical network, which has a relatively small strongly connected core. This is not surprising given the biased patterns observed in the emission of directed messages

degree distribution $P(k)$. The differentiation between homogeneous and heterogeneous networks with respect to their degree is in general associated with the tail of the distribution. If it decays exponentially fast with the degree, we refer to it as homogeneous networks, the most representative example being the Erdös-Rényi (ER) random graph (Erdös and Rényi 1959). On the contrary, when the tail is heavy, one can say that the network is heterogeneous. In particular, scale-free networks are the class of networks where the distribution follows a power law, $P(k) \sim k^{-\gamma}$ the Barabási–Albert model (Barabási and Albert 1999) being the paradigmatic model of this type of graph.

From $P(k)$, we can calculate the moments of the distribution. The n-moment of $P(k)$ is defined as

$$\langle k^n \rangle = \sum_k^N k^n P(k).$$

The first moment $<k>$ is the average degree of the network.

Furthermore, several other measures help to qualify further this categorization. Examples are the average shortest path length $L = <d_{ij}>$, where d_{ij} is the length of the shortest path between node i and node j—very small in complex networks if compared to N—the clustering coefficient C that accounts for the fraction of actual triangles (three vertices forming a loop) over possible triangles in the graph—typically large in social networks, if compared to C in random graphs—and assortativity r (Newman 2002) obtained considering the Pearson correlation coefficient of the degrees at both ends of the edges (assortative behaviour implies that topologically similar nodes link each other, which has interesting interpretations in terms of social structure).

Finally, researchers have not obviated other levels of analysis: Between the global (systemwide) and the microscopic (node) statistical characterization, there exists an intermediate description level: the "meso" scale, in which the relevant analysis unit is a module or community. A modular view of a network offers a coarse-grained perspective in which nodes are classified in subsets on the basis of their topological position and, in particular, the density of connections between and within groups. In social networks, this classification usually overlaps with node attribute data like gender, ideology or professional occupation. To achieve meaningful partitions of complex networks several algorithms have been designed. A thorough review of these techniques by Fortunato appeared in 2010.

3   Data Collection and Modelling

We analyze Twitter activity around the 15M protests for the period comprised between April 25, 2011 at 00:03:26 (20 days before the first mass mobilizations) and May 26, 2011 at 23:59:55 (10 days after the first mass mobilizations and 3 days after the elections). The activity data set follows the posting behaviour of 87,569 users and tracks a total of 581,750 protest messages. Messages related to the protests were identified using a list of 70 *#hashtags* (see Fig. 1 for a tag cloud with the most important ones and Gonzalez-Bailón et al. (2011) for a complete list). The collection of messages is restricted to the Spanish language and to users connected from Spain, and it was archived by a local startup company, Cierzo Development Ltd, using the SMMART Platform. Other SNSs had an important influence on the development of the whole movement. Activity on such sites—as well as blogs, web pages, offline interaction—is blind to this analysis. Admittedly, this work can only account for a partial view of the whole process, a fact that should be kept in mind when reading and interpreting our findings.

An interesting subset of data is obtained when considering only directed messages, i.e. those including mentions, which contain the symbol @. Mentions allow the construction of a *directed* and *weighted* structure, which offers a partial view of the "activity" or "dynamical" network of communication between users. The direction of links indicates a source-target message emission, and link weight stands for the number of times that a user explicitly mentioned another user. Restricting data to messages that contain mentions is useful to characterize direct interactions between pairs of users and the pace at which they get involved with the protests, although two thirds of the information is left aside, i.e. messages that do *not* contain mentions are removed; this was the approach in Borge-Holthoefer

Fig. 1 Word cloud with the main 50 hashtags used to track message activity during the mobilization. Hashtags sizes are proportional to the number of times they appeared in messages, although we have rescaled them to smooth the large difference in their frequencies (which range from 10^5 to 10). A complete list of the 70 hashtags is available in (Gonzalez-Bailón et al. 2011)

et al. (2011). A movie of this dynamic exchange is available online (see http://15m.bifi.es/index_en.php).

In addition to the dynamic network of active mentions, we also reconstructed the network of followers. This offers an almost static view of the relationships between users, although that network also changes over time, its rate of change is significantly smaller than in the dynamic network of mentions (i.e. changes occur in the scale of weeks and months). Data for all the users that were active in the exchange of protest messages were scrapped using the Twitter API and a cloud of 128 machines. The scrap was successful for the 87,569 users identified as active in message exchange, and we obtained their official list of followers. The static network, however, was restricted to those who had some participation in the protests. The resulting structure is a *directed unweighted* network, where direction indicates who follows whom (see also Borge-Holthoefer et al. 2012; Gonzalez-Bailón et al. 2011). This network exhibits a high level of reciprocity: A typical user holds many reciprocal relationships (with other users that are likely to be known personally) plus a few unreciprocated nodes, which typically point at hubs, the so-called network "authorities." Any emitted message from a node i is immediately available to anyone following this user in addition to those mentioned after the @ symbol. This particularity of Twitter is crucial to understand the concept of information cascade in the next sections.

Table 1 summarizes the main topological features of these two views of the data (the static follower network and the activity or dynamical network). Both the static and the dynamic levels reproduce well the "small-world" features (Watts and Strogatz 1998), i.e., low L and high C.

4 Methods and Findings

We paid attention to three levels of analysis when studying the growth and stabilization of the protests: the global dynamics of message exchange over time, the spatial distribution of users at the community level and the recruitment dynamics at the individual level. These three levels provide complementary views of the patterns observed in the data. Details on each of them follow:

4.1 Time Dynamics: Movement Growth and Saturation

The analyses in this section answer the following question: Does a collective mobilization of thousands of actors demand a slow, progressive growth over time or do SNSs enable their abrupt emergence? Figure 2 illustrates the way in which the activity network evolved by gaining adherents. The red squares in the figure represent the proportion of active nodes at time t (with a resolution of 2 h) relative to the total number of users in the network at the end of the growth process; blue circles represent the number of messages produced relative to the total (12 h resolution). Both sets show a similar tendency: The formation of the network and its later increase in size does not proceed in a gradual proportional manner, but rather in a sequence of bursts concentrated in just a few days (from day D to day $D+7$). The process is driven by the offline events surrounding the movement: Data does not control for exposure to offline media, which is likely to have interacted with social influence, or to other sources of information that might have also contributed to the system's growth (like, for instance, offline discussion networks). The number of active users saturates after $D+7$: On May 21 ($D+6$), the day preceding local and regional elections, almost 90 % of the network was already formed.

A key aspect to understand these time dynamics lies at the distribution of degree (static network) and strength (dynamic network). The degree of a node i, k_i, is the number of neighbours it has. The strength s_i of a given node i is defined by the sum of the weights of its links. These magnitudes can further be divided in two measures: On the one hand, we have the in-degree (in-strength) derived from the links incident to the node k_{in} (s_{in}). Conversely, k_{out} (s_{out}) represents the out-degree (out-strength) generated at a node. From these quantities, we can derive $P(k_{in})$ ($P(s_{in})$) and $P(k_{out})$ ($P(s_{out})$), the cumulative probability distributions. $P(k_{in})$ and $P(k_{out})$ are fixed because we assume that the follower network is static, whereas strength distributions can be measured at different instants t of the evolving network.

Figure 3 shows the cumulative distributions of these quantities for several time aggregation windows. As can be seen in the left panels, even before the occurrence of the events that triggered public protests on day D, both $P(s_{in})$ and $P(s_{out})$ follow power laws $P(s) < s^{-\gamma}$, but with different exponents ($\gamma_{in} = 1.1$ and $\gamma_{out} = 2.3$, respectively, as measured at $D+10$). Plots for the degree of the nodes in the follower network (right panels) exhibit similar behaviour, implying the existence of some rare

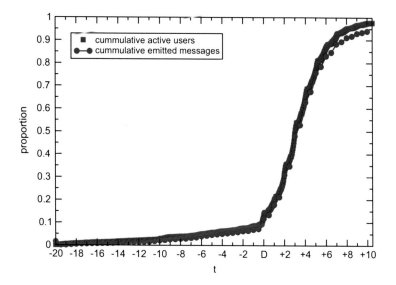

Fig. 2 Temporal evolution of the structure and the activity in the SNS. In *red*, the proportion of nodes that had shown some activity at a certain time *t*. In *blue*, the cumulative proportion of emitted messages as a function of time. Note that the *two lines* evolve in almost the same way

nodes acting as hubs (Barabási and Albert 1999). It is well known that the statistical properties of these variables in other technological, social and natural systems are also heterogeneously distributed. Nonetheless, the fact that the 15M network is scale-free has deep consequences regarding a number of relevant issues, including its origin, complexity, robustness and, from a dynamical point of view, the way in which information flows over the system. As the network obtained comes from the activity of the nodes, the heavy-tailed distribution of both the degrees and strengths of nodes suggests that its dynamics lack any typical or characteristic scale.

Focusing on the left panels of Fig. 3, the dynamic asymmetry between incoming and outgoing degrees or strengths is not surprising either. Indeed, individual behaviour, which ultimately determines the resulting dynamics, is an intended social action, but the emergent properties of the collective behaviour of agents are unintended (Rybski et al. 2009). Essentially, subjects decide when and to whom a given message is sent. Therefore, the aggregate behaviour of all agents and their popularity (i.e., how many mentions a node receives or how many followers it has) result from individual choices. This is what the in-and out-distributions reflect. As a matter of fact, the exponent of the power law characterizing the degree probability distribution $p(s)$ lies in the interval (2,3) as usually found in most real-world networks. Interestingly, spreading dynamics such as rumour and disease propagation processes are most efficient for scale-free networks whose exponent is precisely in this range (Pastor-Satorras and Vespignani 2001; Moreno et al. 2002, 2004). Finally, the strength distribution for the tweets sent, $p(s_{out})$, also resembles a power law function with an exponent larger than 3, although in this case the

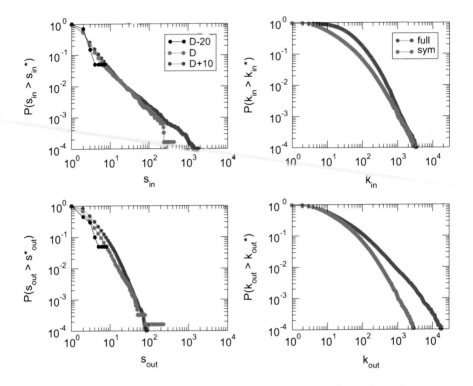

Fig. 3 *Left panels*: Strength distributions for both received (*top*) and sent (*bottom*) messages display a power-law behavior as early as at $D - 2$. The fat-tailed distributions indicate that the 15M activity network is scale-free. The exponents that define the power-laws differ significantly between sent and received messages. *Right panels*: degree distribution of full and symmetric networks, with similar features (see text)

distribution exhibits an exponential cut-off. This might be due to the fact that sending messages has an associated cost in terms of bandwidth availability, the cognitive capacity to produce different messages and ultimately an unavoidable physical limitation to type them (Dunbar 1992, 1993; Gonçalves et al. 2011). Interestingly, the exponential cut-off is mirrored in $P(k_{in})$ (right panel): The capacity to follow other agents in Twitter is cognitively limited, thus the number of users following more than a hundred people on Twitter decays very fast.

One of the main consequences of the functional form of the strength distributions is presented in Fig. 4. The emergence of hubs, which is the signalling feature of scale-free networks, leads to a predictable oligopoly in the way information is spread. In Fig. 4, we observe that the number of tweets sent grows with the number of active users of the network. The curves corresponding to different days (i.e., instances of the dynamic network) nearly collapse into a single one. This means that as users join the network, the traffic generated scales accordingly. The figure indicates that roughly 10 % of active users generated 52 % of the total traffic

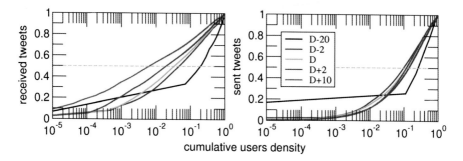

Fig. 4 Information flow: the figure represents the density of tweets received (*left*) and sent (*right*) as a function of the cumulative fraction of active users. For each day, data are normalized by the number of active users at that date. As a reference, the horizontal line corresponds to 50 % of emitted/received tweets. Note that, on $D+10$, less than 1 % of the nodes receive half of the messages. On the contrary, the pattern of tweets sent hardly evolves from the beginning of the movement: 10 % of the active nodes produce 50 % of the messages. This asymmetry is coherent with the differences observed for the strength distributions

[which is another indication of the dynamical robustness of the network to random failures but, at the same time, of its fragility to attacks directed towards that 10 % of users (Albert et al. 2000; Cohen et al. 2000)]. These patterns, however, are in sharp contrast with the activity patterns that correspond to received tweets. In this latter case, the number of in-strength hubs decreases with time. As shown in the figure, by $D+10$, less than 1 % of users receive more than 50 % of the information. These nodes are identified as main receptors (government) or as potential spreaders (mass media) of messages. However, what at a priori seems to be a good choice turns out to be harmful for the process of information spreading. These hubs, which we call *information sinks*, do receive a lot of messages but rarely ever act as spreaders, which means they do not pass those messages along to their followers. These asymmetries in the information flow critically shape other global processes like information cascades, which, as will be shown below, are very difficult to trigger.

4.2 Spatial Dynamics: Communities and Geography

A clustered or modular structure is pervasive in many natural, social and techno-logical networks. Modules are islands of highly connected nodes separated by a relatively small number of links, and in social networks, they are created by the fact that actors tend to gather with those who share similar attributes like cultural traits or professional interests (Arenas et al. 2004; Lozano et al. 2008; Blondel et al. 2008; Fortunato 2010). In political networks, homophily clusters actors around ideologies or opinions (Adamic and Glance 2005; Conover et al. 2011).

We analyzed the community structure of the 15M activity network at $t = D + 10$ (that is, as its size is stabilized). We applied a random walk-based clustering algorithm that optimizes a map equation on a network structure (Rosvall and Bergstrom 2008). Although alternative community detection algorithms are at hand, we choose the previous strategy because it is suited for networks in which the dynamics of information flow is relevant (Pons and Latapy 2005; Rosvall and Bergstrom 2008), as is our case. The output of the algorithm is a partition made up of 6,388 modules. Most of these communities have less than ten nodes, so we focus our analysis on the 30 most important modules from a dynamic perspective, i.e. those which concentrate most of the random walker's activity. These modules do not necessarily coincide with the first 30 communities ranked according to their size, but all of them contain over 100 nodes, accounting for $\sim$15 % of the users. Figure 5 shows these 30 communities in a collapsed view (each node represents a community) (Edler and Rosvall, http://www.mapequation.org). Each community is assigned a tag, which corresponds to the most central node in that community. These nodes play an outstanding role in the dynamics of information: Modules are highly hierarchical, and nodes that are central in their communities (i.e. the local hubs) are hubs in a global scale as well.

This community structure sheds new light into the social dimension of the protests. First, tags identifying the 30 largest communities are highly heterogeneous. Six of these modules correspond to important mass media (newspapers and television), suggesting that users rely on these agents to amplify their opinion. The same can be said of three modules corresponding to famous journalists. More interestingly, seven modules correspond to on-line activists and/or veteran bloggers. These agents are unknown to most people, but they are present in the network from its birth and enjoy a solid reputation that facilitates their being considered a reference in the movement. Remarkably, seven modules are formed by camps in seven different cities. Madrid is of course the main one, as the movement began there (#acampadasol, which comprehends over 3,000 nodes). Other cities are Barcelona, Granada, Zaragoza, Valencia, Seville, and Pamplona.

The fact that communities are heterogeneously dominated (either by a person, an organization or a place) suggests some interesting conclusions. First, unlike previous works (Adamic and Glance 2005; Conover et al. 2011), modules are not homogeneously defined (e.g., groups corresponding to political tendencies). Instead, the activity network breaks down into groups that are representative of different actors in the process of mobilization (mass media, opinion groups, journalists, etc.). Second, communities reflect the relative autonomy of each of the assemblies throughout the Spanish geography: Each of these modules hardly connects to any other, indicating a low communication between them. Third, Madrid, the capital, poses one exception to this pattern: Minor camps hold a strong communication interchange with the community represented by the capital hashtag #acampadasol. In conclusion, these analyses show that the movement is highly centralized because most peripheral communities are only influenced by Madrid.

From a geographical point of view, the analyses suggest that in spite of the potential that Web 2.0 technologies have to bridge geographical distance, they are

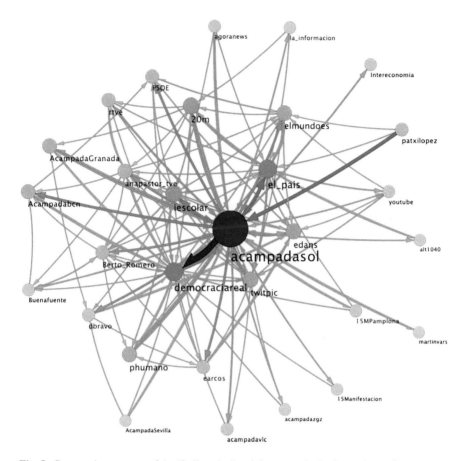

Fig. 5 Community structure of the "Indignados" activity network: the figure shows, in a compact view in which each node represents a community, the 30 most important modules. They can be identified by a single node, around which the community is organized. These *local hubs* (labeled in the figure) agglutinate modules and act as information bridges connecting the whole network

mostly used to communicate with geographically close people. In other words, the network is *global*, but communication is mainly *local*. This is further verified in Table 2, where we have summarized the percentage of people whose public geolocation information (when signing up for a Twitter account) coincides with that of the community represented by the city hashtag. Admittedly, location at sign-up time is not free of error—people might move or simply be inexact when delivering that information. Nonetheless, data suggest that geolocation is a significant factor in users' interaction.

Table 2 Geographic origin of nodes in region-based communities

Community tag	Area	Fraction of users from same area (%)
@acampadasol	Madrid	54
@acampadabcn	Barcelona	81
@acampadavlc	Valencia	63
@acampadazgz	Zaragoza	82
@acampadagranada	Granada	53
@acampadasevilla	Seville	83
@15MPamplona	Pamplona	71

Region-based modules are mostly formed by nodes whose geographical origin coincides with that of the most central node in the community. This statement is clear for almost all these modules, except in the case of Madrid and Granada. The case of Madrid is not surprising, given that #acampadasol is the reference of the whole movement, thus the community organized around this actant is a more heterogeneous one. Granada is a more intriguing exception

4.3 Recruitment Dynamics: The Activation of Users

The third level of analyses focuses on the mechanisms that prompted users to start sending protest messages. In particular, we analyze processes of recruitment (when and how do users join this instance of collective action?) and of information diffusion (how does the system help share—or filter out—information?).

4.3.1 The Network Position of Recruitment Seeds

The analysis conducted to answer the first question provides deeper insights into why horizontal organizations (like the online-born platform coordinating the protest, see Appendix) are so successful at mobilizing people through SNSs: Their decentralized structure plants activation seeds randomly at the start of the recruitment process, all over the network structure and in the local contexts created by a myriad users and organizations. Users who start sending messages before anyone else are the leaders of the movement, but they are embedded in very different local networks: Some are more central, some more peripheral; their diverse positions within the overall network start activation chains that end up percolating to the entire network.

Time-stamped data tell us the exact moment when users start emitting messages and allow us to distinguish between activists leading the protests and those who reacted in later stages. It also allows us to calculate, for every user, how many of their contacts had already sent protest messages at the time of their activation (k_a/k_{in}).

Activists with an intrinsic willingness to participate have a threshold $k_a/k_{in} \approx 0$, whereas those who need a lot of pressure from their local networks before they decide to join are in the opposite extreme $k_a/k_{in} \approx 1$ (see Gonzalez-Bailón et al. 2011 for details).

In order to identify the network position of recruitment seeds, we used the k-shell decomposition (Alvarez-Hamelin et al. 2006, 2008) to dissect the network in

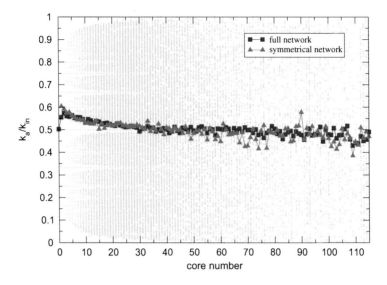

Fig. 6 Activation time as a function of the core of nodes: early adopters, responsible for the recruitment process, are not located at privileged network positions. On the contrary, they are spread all over the topology. The maximum core in the static network is $k_S = 141$

centrality layers where we then locate the leaders of the movement. The k-shell decomposition assigns a shell index k_S or "core" to each user by pruning the network down to users with more than k neighbours. The process starts removing all nodes with degree $k = 1$, which are classified (together with their links) in a shell with index $k_S = 1$. Nodes in the next shell, with degree $k = 2$, are then removed and assigned to $k_S = 2$ and so forth until all nodes are removed (and all users are classified). Shells are layers of centrality in the network: Users classified in shells with higher indexes are located at the core, whereas users with lower indexes define the periphery of the network. k_S in the static network ranges from 1 to 141, which means that core users have at least 141 connections each.

Figure 6 combines the activation time with the core of each user in the static network. The overall spread of gray dots and the averages depicted for each core clearly show that any node, regardless of the k-shell it belongs to, can be activated at any time. This result confirms that recruitment takes place through random and distributed seeding.

4.3.2 The Network Position of Information Spreaders

Information diffusion tells us not how the system attains a critical mass, but how it maintains its activity. In the context of collective action, information diffusion plays a key role to coordinate action and to keep adherents informed and motivated. Understanding the dynamics of such diffusion is important to identify the users that are more likely to transform the emission of a single message into a global information cascade.

An information cascade starting at a trigger occurs whenever a piece of information is (more or less unchanged) repeatedly forwarded towards other users. If one of those who "hear" the piece of information decides to forward it, he becomes a spreader, otherwise he remains as a mere listener. The information cascade becomes global if the final number of affected users N_C (including the set of spreaders and listeners plus the seed) is comparable to the size of the whole system N. Intuitively, the success of an information cascade should depend on whether spreaders have a large set of followers or not (Fig. 7). This fact highlights the entanglement between dynamics and the underlying (static) structure.

Our definition of cascade takes time into consideration assuming that regardless of the exact content of a message, two nodes belong to the same cascade as consecutive spreaders if they are connected (i.e. the latter follows the former) and they show activity within a short time interval Δt (Borge-Holthoefer et al. 2012). The probability that exogenous factors are leading activation is in this way minimized. This operationalization of cascades borrows the concept of spike train from neuroscience, i.e. a series of discrete action potentials from a neuron taken as a time series.

We apply the latter definition to explore the occurrence of information cascades in the data. In practice, we take a seed message posted by i at time t_0 and mark all of i's followers as listeners. We then check whether any of these listeners showed some activity at time $t_0 + \Delta t$. This is done recursively until no other follower shows activity. Figure 7 gives an illustrative summary of the method we followed to reconstruct cascades. In our scheme, a node can only belong to one cascade. We distinguish information cascades (or just cascades for short) from spreader cascades. In information cascades, we count any affected user (listeners and spreaders), whereas in spreader cascades, only spreaders are taken into account.

We measure the size distribution of cascades and spreader cascades for three different scenarios: one in which the information volume is low (slow-growth phase, from $D - 20$ to $D - 10$), one in which activity is very high (explosive phase, $D - 2$ to $D + 6$) and one that considers all available data (which spans a whole month and includes the two previous scenarios plus the time in between, $D - 20$ to $D + 10$). Within the different time periods—slow growth, explosive phase and complete time-span—different time windows have been set to assess the robustness of our results.

Our proposed scheme relies on the contagious effect of activity; large time windows, i.e., $\Delta t > 24$ h, are not considered.

The upper panels (a, b, c) of Fig. 8 reflect that an information cascade of the size of the system can be reached in any of the three phases. As expected, these large cascades are always rare events, as the power law probability distributions point out. This result is in perfect accordance with our analysis at the global scale (see Figs. 3 and 4) and robust to different temporal windows up to 24 h. In contrast, lower (d, e, f) panels do show significant differences between periods. Specifically, the distribution of involved spreaders in the different scenarios changes radically

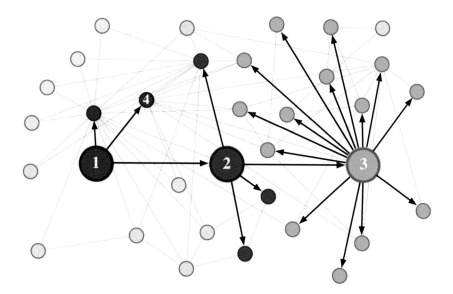

Fig. 7 The figure illustrates the concept of cascade that is used throughout this article. Times at which spreaders emit a message is color-coded. Spreaders are emphasized as larger circles. User 1 emits a message at time t, and all of his followers automatically receive it. Thus, they are already counted as part of the cascade (*red circles*). One of his followers (user 2), driven by the previous message, decides himself to participate at time $t + \Delta t$, posting a message himself. A second set of followers is included in the cascade. Finally, a third node (user 3) joins in and spreads the cascade further at time $t + 2\Delta t$. A node cannot be counted twice, note for example that user 4 is also following node 3. Many nodes remain unaffected, because they are not connected to any of the spreaders. The final size of the cascade is $N_C/N = 22/34$; the success of the cascade largely depends on the capacity to contact a "leader" or "privileged spreader," i.e., a hub to whom many people listens and who decides to participate. The interesting point, however, is that the number of spreaders needed to attain such success is very low (3), and over 50 % of the cascade is triggered by just one of them. Adapted from Borge-Holthoefer et al. (2012)

from the explosive period (Fig. 8f) to the slow-growth period (Fig. 8d); the distribution that considers the whole period of study just reflects that the explosive period (in which most of the activity takes place) dominates the statistics. What these differences suggest is that to attain similar results (a systemwide cascade), a proportionally much smaller amount of spreaders (users who receive a message and pass it on) is needed in the slow-growth period.

To identify the network position of these spreaders, we capitalize on previous work suggesting that centrality (measured as the k-core, see previous section) enhances the capacity of a node to be key in disease (Kitsak et al. 2010) and rumour (Borge-Holthoefer and Moreno 2012) spreading processes. In the work by Kitsak and colleagues (2010), it is discussed whether the degree of a node (its total number of neighbours, k) or its k-core (a centrality measure) can better predict the spreading capabilities of such a node. Note that the k-shell decomposition splits a network in a

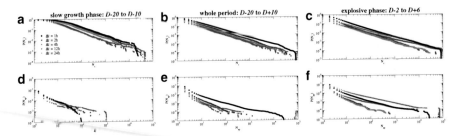

Fig. 8 *Upper panels* (**a**, **b**, **c**): Cascade size probability distributions for the different periods considered. *Lower panels* (**d**, **e**, **f**): Probability distributions of spreaders involved in the cascades for the same periods. The exact periods considered in the analyses are indicated at the top of each panel. Adapted from Borge-Holthoefer et al. (2012)

few levels (over a hundred), while node degrees can range from one or two up to several thousands.

Figure 9 explores the same idea, but in relation to information cascades. On the left is a circular layout of the follower network. Central nodes (high k-core) are placed in the centre. Nodes are coloured according to their "cascading capacity," i.e. the size of a cascade triggered in a specific node. The upper central panel of this figure shows the spreading capabilities as a function of classes of k-cores. Specifically, we take the seed of each particular cascade and save its coreness and the final size of the cascade it triggers. Having done so for each cascade, we can average the success of cascades for a given core number. Remarkably, for every phase under consideration (slow growth, explosive and full), a higher core number yields larger cascades. Exactly the same conclusion (and even more pronounced) can be drawn when considering degrees (lower central panel).

These results suggest that both degree and k-core are good cascade size predictors. However, the interest of the histograms in Fig. 9 lies in the high-end regions: While there are a few hundred nodes in the high cores (and even over a thousand in the last core), the highest degrees account only for a few dozen nodes. In practice, this means that only extremely high degrees, which are very rare, can produce large cascades. On the contrary, high cascading capabilities are distributed over a wider range of cores, which in turn contain a significant number of nodes.

The previous findings show that joint dynamics of recruitment and coordination took place in this mobilization. Early adopters in the first days of the protests acted as recruiters and ultimately reached a critical mass of core users. Users at this core in turn contributed to the growth of the movement by generating cascades of messages that triggered new activations. These joint dynamics illustrate why Twitter has played a prominent role in so many recent protests and mobilizations: It combines the global reach of broadcasters with local, more personalized relations; in the light of our data, both features are important to articulate the growth of a movement. These features, however, are necessary, not sufficient, conditions: Offline events and offline information diffusion also contributed to the growth of a protest that we only track online.

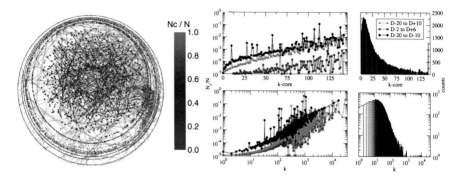

Fig. 9 *Left panel*: nodes in the static network arranged according to their *k*-core: higher *k*-cores indicate that nodes are more central, node size accounts for degree centrality, and node color indicates the maximum size of the cascades generated by the user (users generating the largest cascades are depicted in *orange*). *Central upper panel*: average spreading capacity (with respect to the system size) of nodes grouped according to their *k*-core. N_C/N grows with coreness, but the bursty period (*red squares*) evidences a much less clear tendency, with many fluctuations and a lower overall spreading capacity if compared to the relaxed period (*black circles*). *Central lower panel*: The same information is showed as a function of the degree. Again, the slow growth period is the best one at predicting the extent of a cascade. Interestingly, average cascades for highest degrees outperform those triggered by highest *k*-core nodes by an order of magnitude. See main text for discussion on this aspect. *Right panels* show the *k*-core and degree distributions, i.e., how many nodes belong to each class. Note that the highest core contains over 1,000 users

5   Conclusions

It is generally accepted that social networking sites are dramatically altering the way in which humans communicate. Such platforms offer a public domain where political opinion and social activism emerge, creating in the process data that offer scholars the opportunity to test old theoretical claims of how political and social protests emerge. This unprecedented volume of data, however, demands modern tools and research strategies to tackle its complexity from different levels. This chapter has provided one example of the insights that following such a research avenue can yield.

A network perspective of mass mobilizations, as observed from online activity, sheds new light on the old problem of collective action. We analyzed one instance of online mobilization integrating three different levels of analysis: time dynamics, spatial organization and recruitment mechanisms. We paid close attention to the distinction between dynamic activity (i.e. the exchange of messages) and the underlying structure supporting that activity (i.e. the following/follower connections of Twitter users). The analysis of time dynamics reveals the global characteristics of the "Indignados" movement, allowing us to describe the sudden growth of the protests after a relatively long period of slow brewing. This level of analysis provides also some insights into how information flows and the high influence that a minority of users have on global dynamics. The modular structure of the network formed by the exchange of messages reveals that groups of users gather around relatively autonomous entities like mass media, famous journalists or city-based camps. A somewhat surprising finding is that in spite of the global reach of SNSs like Twitter, users mainly employ this tool to communicate locally, mostly to coordinate actions that were taking place offline. Finally, the analysis of recruitment patterns and information dynamics, which demands a finer-grained view of the 15M movement, reveals the importance that local contexts have to understand activity at the individual level. For the movement to attain a critical mass of adherents, recruiters planted seeds all through the network, but once the system percolates, only those users who lie at the core are influential enough so as to trigger large information cascades, which make global coordination possible.

These findings provide a comprehensive view of one particular instance of emergent behaviour integrating different levels of analysis to uncover the mechanisms underlying collective action. However, far from exhausting the subject, this research generates new questions. A fundamental problem with online data is that it only offers insight into one side of the story, excluding from the focus of analysis a number of relevant agents (like, for instance, traditional mass media) that also influence the process under study. Political activism, in particular, is a collective endeavour with manifestations at many levels ranging from public demonstrations and camps to the presence in news and mass media reporting. Thus, the offline events that also contributed to trigger the growth of the movement remain unknown and are beyond the scope of this work. Another important limitation relates to the content of the messages exchanged: What type of content is more likely to be disseminated on a global scale? While theoretical models of information spreading have long been studied and modelled, we are still trying to understand how the attributes of the information exchanged (for instance, its emotional content and appeal) is related to its diffusion. Finally, there are many determinants of geographical propagation that remain unexplored: The spatial location of users contributing to online activity is relevant to understand part of the dynamics, but this does not illuminate the factors that made original protesters appear in certain places and not others. Our study locates the origins of the protest in an online network, but cannot answer the question of why the protest flourished at a time or place.

In spite of all these limitations, however, there is much to be gained from analyzing activity in online networks. Not only is human activity increasingly shifting to that digital realm demanding new ways to understand social dynamics and behaviour; online networks are also encouraging collaboration across disciplines (complexity science, physics, sociology) that were mostly unconnected before, promoting the development of novel methods and theoretical frameworks. Digital data open new windows to analyze long elusive processes like contagion and social diffusion and the way in which they interact with overlapping network structures (like communication and spatial networks). Better theories and models will emerge from that knowledge, but also better tools for political participation.

Acknowledgements J.B-H is partially supported by the Spanish MICINN through project FIS200801240. S.G-B. is partially supported by the Spanish MICINN projects CSO2009-09890 and CSD2010-00034. Y. M. is supported by the Spanish MICINN through projects FIS2008-01240 and FIS2009-13364-C02-01 and by the Government of Aragon (DGA) through the grant No. PI038/08.

Appendix: Protests Organizers

The "Indignados" movement is a civic initiative with no party or union affiliation that emerged as a reaction to perceived political alienation and to demand better channels for democratic representation. The first mass demonstration, held on Sunday, May 15, was conceived as a protest against bipartidism and the management of the economy in the aftermath of the financial crisis. It was organized by the digitally coordinated platform "Democracia Real Ya" ("Real Democracy Now"), born online about 3 months before the first day of demonstrations. Hundreds of entities joined the platform, from small local associations to territorial delegations of larger groups like ATTAC (an international anti-globalization organization) or "Ecologistas en Acción" ("Ecologists in Action"). Signatories of the original call included student associations, bloggers, defenders of human rights and people from the arts, but also hundreds of individual citizens of different ages and ideologies. Under the motto "toma la calle" ("take the streets"), the movement organized peaceful protests that brought tens of thousands of people to the streets of more than 50 cities all over the country, with Madrid and Barcelona leading in numbers.

References

Adamic L, Glance N (2005) The political blogosphere and the 2004 us election: divided they blog. In: Proceedings of the 3rd international workshop on link discovery. ACM, pp 36–43
Albert R, Barabási A (2002) Statistical mechanics of complex networks. Rev Mod Phys 74(1):47–97
Albert R, Jeong A, Barabási A (2000) Error and attack tolerance of complex networks. Nature 406:378–382
Alvarez-Hamelin J, Dall Asta L, Barrat A, Vespignani A (2006) Large scale networks fingerprinting and visualization using the k-core decomposition. Adv Neural Inf Process Syst 18:41
Alvarez-Hamelin J, Dall'Asta L, Barrat A, Vespignani A (2008) k-core decomposition of Internet graphs: hierarchies, self-similarity and measurement biases. Netw Heterogeneous Media 3(2):371–393
Arenas A, Danon L, Díaz-Guilera A, Gleiser PM, Guimerà R (2004) Community analysis in social networks. Eur Phys J B 38:373–380
Bakshy E, Hofman J, Mason W, Watts D (2011) Everyone's an influencer: quantifying influence on twitter. In: Proceedings of the fourth ACM international conference on Web search and data mining. ACM, pp 65–74
Barabási A (2005) The origin of bursts and heavy tails in human dynamics. Nature 435 (7063):1251

Barabási A, Albert R (1999) Emergence of scaling in random networks. Science 286:509

Blondel V, Guillaume J, Lambiotte R, Lefebvre E (2008) Fast unfolding of communities in large networks. J Stat Mech P10008

Boccaletti S, Latora V, Moreno Y, Chavez M, Hwang D (2006) Complex networks: structure and dynamics. Phys Rep 424(4–5):175–308

Borge-Holthoefer J, Moreno Y (2012) Absence of influential spreaders in rumor dynamics. Phys Rev E 85:026116

Borge-Holthoefer J, Rivero A, García I, Cauhé E, Ferrer A, Ferrer D, Francos D, Íñiguez D, Pérez M, Ruiz G et al (2011) Structural and dynamical patterns on online social networks: the Spanish May 15th movement as a case study. PLoS One 6(8):e23883

Borge-Holthoefer J, Rivero A, Moreno Y (2012) Locating privileged spreaders on an online social network. Phys Rev E 85:066123

Cohen R, Erez K, ben Avraham D, Havlin S (2000) Resilience of the internet to random breakdowns. Phys Rev Lett 85(21):4626–4628

Conover M, Ratkiewicz J, Francisco M, Gonçalves B, Flammini A, Menczer F (2011) Political polarization on twitter. In: Proceedings of the 5th international conference on weblogs and social media

Dorogovtsev S, Goltsev A, Mendes J (2008) Critical phenomena in complex networks. Rev Mod Phys 80(4):1275–1335

Dunbar R (1992) Neocortex size as a constraint on group size in primates. J Hum Evol 22 (6):469–493

Dunbar R (1993) Coevolution of neocortical size, group size and language in humans. Behav Brain Sci 16(4):681–693

Edler D, Rosvall M. The map generator software package. http://www.mapequation.org

Erdös P, Rényi A (1959) On random graphs. Publ Math (Debrecen) 6:290–297

Fortunato S (2010) Community detection in graphs. Phys Rep 486(3–5):75–174

Freeman L (1977) A set of measures of centrality based upon betweenness. Sociometry 40:35–41

Gonçalves B, Perra N, Vespignani A (2011) Modeling users' activity on twitter networks: validation of Dunbar's number. PLoS One 6(8):e22656

Gonzalez-Bailón S, Borge-Holthoefer J, Rivero A, Moreno Y (2011) The dynamics of protest recruitment through an online network. Sci Rep 1:197

Granovetter M (1973) The strength of weak ties. Am J Sociol 78:1360–1380

Kitsak M, Gallos L, Havlin S, Liljeros F, Muchnik L, Stanley H, Makse H (2010) Identification of influential spreaders in complex networks. Nat Phys 6(11):888–893

Kwak H, Lee C, Park H, Moon S (2010) What is twitter, a social network or a news media? In: Proceedings of the 19th international conference on World Wide Web. ACM, pp 591–600

Lozano S, Arenas A, Sánchez A (2008) Community connectivity and heterogeneity: clues and insights on cooperation on social networks. J Econ Interac Coord 3(2):183–199

Milgram S (1967) The small world problem. Psychol Today 2:60–67

Moreno Y, Pastor-Satorras R, Vespignani A (2002) Epidemic outbreaks in complex heterogeneous networks. Eur Phys J B 26(4):521–529

Moreno Y, Nekovee M, Vespignani A (2004) Efficiency and reliability of epidemic data dissemination in complex networks. Phys Rev E 69(5):055101

Newman M (2002) Assortative mixing in networks. Phys Rev Lett 89(20):208701

Newman MEJ (2003) The structure and function of complex networks. SIAM Rev 45:167–256

Onnela J, Saramäki J, Hyvönen J, Szabó G, Lazer D, Kaski K, Kertész J, Barabási A (2007) Structure and tie strengths in mobile communication networks. Proc Natl Acad Sci 104 (18):7332

Pastor-Satorras R, Vespignani A (2001) Epidemic spreading in scale-free networks. Phys Rev Lett 86(14):3200–3203

Pons P, Latapy M (2005) Computing communities in large networks using random walks. Lect Notes Comput Sci 3733:284

Rapoport A (1953) Spread of information through a population with socio-structural bias: I. Assumption of transitivity. Bull Math Biophys 15:523–533

Ratkiewicz J, Fortunato S, Flammini A, Menczer F, Vespignani A (2010) Characterizing and modeling the dynamics of online popularity. Phys Rev Lett 105(15):158701

Rosvall M, Bergstrom C (2008) Maps of random walks on complex networks reveal community structure. Proc Natl Acad Sci 105(4):1118

Rybski D, Buldyrev S, Havlin S, Liljeros F, Makse H (2009) Scaling laws of human interaction activity. Proc Natl Acad Sci 106(31):12640–12645

Sun E, Rosenn I, Marlow C, Lento T (2009) Gesundheit! Modeling contagion through facebook news feed. In: Proceedings of the ICWSM 9

Travers J, Milgram S (1969) An experimental study of the small world problem. Sociometry 32(4):425–443

Watts D, Strogatz S (1998) Collective dynamics of 'small-world' networks. Nature 393:440

The Strength of Tweet Ties

Rob Schroeder, Sean F. Everton, and Russell Shepherd

Abstract While in 2011 protesters took to the streets and gathered in Tahrir Square in Cairo, people in Egypt and across the world started discussing the protests on Twitter. With its short and simple messages, Twitter turned out to be an effective venue for diffusion of ideas and opinions. This chapter explores how social movements and other forms of collective action may be able to use Twitter to frame grievances in ways that resonate with their target audience. In particular, using Twitter data collected during the protests in Egypt, this paper examines whether the use of Twitter by Egyptian activists helped to diffuse the Arab Spring frame across Egypt and generate greater social cohesion around their messages. Our results lend tentative support to the hypothesis that it in fact did. When activists or members of the traditional media are in positions of brokerage, the level of cohesion within a community is greater than it would be otherwise. That highly central activists have the largest effect suggests that because their position within the network allows them to broker the flow of information, they are able to use Twitter to frame events in ways that resonate with others.

1 Introduction

Around 11:30 in the morning on December 17, 2010, a young man walked in front of the governor's office in a rural Tunisian town, dowsed himself with gasoline purchased from a nearby gas station, and lit himself on fire. Earlier that day, police officers had beat 26 year-old Mohamed Bouazizi in the street for selling fruit without a permit—his sole means of supporting his family. In the aftermath, it

---

R. Schroeder (✉) • S.F. Everton • R. Shepherd
Naval Postgraduate School, Monterey, CA, USA
e-mail: rcschroe@nps.edu

N. Agarwal et al. (eds.), *Online Collective Action*, Lecture Notes in Social Networks, DOI 10.1007/978-3-7091-1340-0_10, © Springer-Verlag Wien 2014

became apparent that Bouazizi simply lacked the money necessary to bribe a local official, who responded by having her aides rough him up and confiscate his wares. Mohamed's quest to retrieve his scales and cart took him the local governor's office, where his complaints fell on deaf ears. Fed up with the corruption, unable to find work in a town with a 30 % unemployment rate, his applications rejected by the army and the government, Bouazizi caught the peoples' attention with an extreme act of protest. What happened next shocked everyone in Tunisia, across North Africa and even around the world. Protests sprung up from town to town, such that by the time Mohamed died on January 4th, the entire country was engulfed by outrage at the government. That is the story of how a theretofore unknown town in rural Tunisia became birthplace to a revolution.

As the story of protests spilled out across news services, satellite feeds, and fiber optic cables, people across North Africa began identifying with the popular sentiment of what was quickly becoming a full-scale revolution in Tunisia. In addition to protests at home, Bouazizi's suicide coalesced individuals to recognizing that they too had similar grievances. These individuals acted on those grievances by actively protesting against the government in public spaces as well as sharing their experiences and protesting in the digital space (Howard and Hussain 2011). Information about these protests spread across North Africa and Middle East, including Egypt where the grievances resonated across various opposition movements. Similar to Tunisia, these Egyptian opposition movements used social media to congregate in the digital realm where connections between previously disconnected groups were brokered, shared grievances among these groups and activists, and spread their messages around the world (Lim 2012).

The protest model that emerged was primarily comprised of young protestors and organized through extensive use of online communications (particularly social media); it garnered widespread international attention and united previously disparate groups under the larger anti *status quo* movement. There is a continuing debate on the ability of social media to affect states more generally, not just in the context of the Arab Spring as applied here. On the one hand, social media enthusiasts point toward the success of the revolutions in Tunisia and Egypt as evidence of social media's power. They argue that because Internet-based communications operate independent of the state apparatus and external to the concept of the nation-state, this borderless public forum allows for debate across political, ethnic, religious and class lines, essentially nullifying the divide-and-conquer strategies of many authoritarian regimes. Conversely, skeptics point to the inability of Internet communications to escape well-established methods of state control and manipulation, such as how the Egyptian regime cut off Internet access in response to early protests, although this did have the unintended consequence of driving even more people to the street.

If social media did play a decisive role in fomenting the protests in Egypt, how did it do so? Lynch (2011) observes there are four ways through which social media could affect contemporary Arab states, "(1) promoting continuous collective action; (2) limiting or enhancing the mechanisms of state repression; (3) affecting international support for the regime; and (4) affecting the overall control of the public sphere." Although these are all interrelated, this chapter focuses on the dynamics between social media and revolution via Lynch's fourth channel. Using social network data extracted from Twitter messages (i.e., "tweets") related to the 2011 Egyptian Revolution, we examine how this social network changed and evolved in ways that helped frame the grievances of the Egyptian people in such a way that they helped give birth to the revolution. We nest our evaluation of social media's role within the context of contemporary social movement theory (in particular, the collective process generally referred to as framing), in order to illuminate the underlying structural causes of Twitter's role in modern activism. To be clear, we do not argue that social media tools were all that activists needed in order to mobilize the Egyptian people. Rather, we see it as one tool among many that activists used to frame and spread their revolt against the Mubarak regime. The remainder of the chapter proceeds as follows: We begin with a very brief overview of the events leading up to and including the Egyptian revolution. We then turn to a discussion of social movement theory, focusing primarily on the role that the framing of grievances plays in a social movement's emergence; we then consider how social media tools, such as Twitter, can possibly facilitate the framing process. Next, we present our data and the methods we use to analyze them before examining the results of our analysis. This section is followed with a brief reflection on the implications of our findings.

2 Background

On January 25th, demonstrators mounted a protest in opposition to Hosni Mubarak's government in Egypt that ultimately led to him to step down from power. This event was just one in a number of popular protests that led to the collapse of authoritarian regimes across North Africa that collectively have been labeled the Arab Spring. These uprisings were not entirely surprising. Since the turn of the century, political demonstrations have massed in the streets of North African and Middle Eastern countries numerous times, such as the demonstrations in 2000 that supported the second Palestinian intifada, the ones in 2002 that protested Israeli operations in the West Bank, and those in 2003 that protested the Iraq War (International Crisis Group 2011). Political demonstrations against Mubarak were also not new. Take, for instance, the Kefaya movement, which emerged in 2004 in response to changes in the Egyptian constitution that allowed Mubarak to remain in power for a fifth term. It successfully organized numerous demonstrations in 2005 and 2006 and united a number of political parties holding a wide range of different ideologies—communists, nationalists, Islamists—under the banner of regime change (Oweidat et al. 2008). More pertinent to our purposes here, Kefaya utilized new information technology in pursuit of its goals. For example, it created a website that allowed users to post grievances online, it regularly communicated with members and others via electronic messages, it posted advertisements online and in independent media outlets, it published banners and political cartoons on the

Internet, it gathered and distributed video and photographic evidence that documented sexual and physical harassment by state security officers, and so on (Oweidat et al. 2008). Although the Kefaya movement eventually disbanded, other movements adopted portions of its organizational structure as well as its methods of using new technology.

Two examples of this are the April 6th Youth movement, which was formed to support striking textile workers, and the "We Are all Khaled Said" movement, which began in response to the death of an Alexandrian man, who allegedly died at the hands of Egyptian police. Shortly after the Tunisian movement swept President Zine el Abidine Ben Ali from power, activists from these two movements (as well as others) created a Facebook page that called for a large scale protest on January 25th, the National Egyptian Police Day holiday. This initial protest gained large scale support from various groups of differing ideologies and resulted in demonstrations across the country, where protesters often outnumbered police and led those protesting in Cairo to converge on Tahrir Square (International Crisis Group 2011).[1] As the protests continued, the movement picked up more and more supporters that helped enable it to resist attempts by the Mubarak regime to shut it down, and on February 11th, Mubarak was forced to resign.[2] As noted earlier, during the Egyptian protests individuals used Twitter and other social media tools to spread the word about the protests, not only inside Egypt but across the world (Salama 2011). The questions we take up in this chapter, however, are whether social media played a decisive role in fomenting the Egyptian protests, and if it did, how? To begin to answer these questions, we turn to consideration of social movement theory, the process of framing, and how social media tools, such as Twitter, might play a role in the framing of grievances.

3   Social Movement Theory, Framing, and Twitter

It is often assumed that all that is required for a social movement, an insurgency, or other form of collective action to emerge is for enough individuals to become sufficiently angry about a particular social condition or set of conditions. There is an element of truth in this assumption. Grievances are generally necessary in order for collective action to occur. However, they are not enough. As social movement scholars have noted, while in most societies there are plenty of individuals who are dissatisfied with the status quo, few become activists, form a social movement, or engage in contentious politics (McCarthy and Zald 1977). Instead, other factors

---

[1] Tahrir Square is a major public square in downtown Cairo. It is formerly the site of British barracks, and after the British left Egypt in 1949, King Farouk raised the Egyptian flag and renamed the square Tahrir "Liberation" square. It has also been the site of numerous protests and marches prior to the ones in January 2011 (International Crisis Group 2011).

[2] For more information on the Egyptian demonstrations, see International Crisis Group (2011).

need to fall into place before a disaffected group of individuals is able to mobilize successfully (McAdam 1982; McAdam and Snow 2010; McAdam et al. 1988, 2001).

Not only do people need to harbor grievances of some kind, but (1) the grievances have to be framed in such a way that people recognize they share them with others and believe that collectively they can do something about them (i.e., *insurgent consciousness*), (2) the disaffected population needs to have access to and be able appropriate sufficient resources (i.e., *sufficient mobilizing resources*), and (3) the group needs to perceive (either correctly or incorrectly) that the broader sociopolitical environment is vulnerable to collective action or that it represents a significant threat to the group's interests or survival (i.e., *expanding opportunities or increased threats*) (McAdam and Snow 2010; McAdam et al. 2001). Figure 1 illustrates the interaction of these factors. In isolation, none of them is sufficient to generate and sustain an insurgency. Together, however, they make collective action more likely (Smith 1991: 64–65).[3]

As one might guess, however, the development of an insurgent consciousness is anything but automatic, and scholars have devoted considerable attention to the process by which movements frame grievances in ways that resonate with their target audience so that they are compelled to act in order to alter the social situation:

> The framing perspective on collective action and social movements views movements not merely as carriers of existing ideas and meanings, but as signifying agents actively engaged in producing and maintaining meaning for constituents, antagonists, and bystanders... The verb *framing* is used to conceptualize this signifying work, which is one of the activities that social movement leaders and their adherents do on a regular basis (Snow and Byrd 2007: 123–124).

David Snow and his colleagues (Snow and Benford 1988; Snow and Byrd 2007; Snow et al. 1986) have identified three stages in the framing process: diagnostic, prognostic, and motivational (Snow and Benford 1988; Snow and Byrd 2007). The first stage, diagnostic framing, focuses on questions such as, "What's the problem?" "What went wrong?" and "Who or what is to blame?" Answers to these questions and others can "recast features of political or social life that were previously seen as misfortunes or unpleasant but tolerable facts as intolerable injustices or abominations that demand transformation" (Snow and Byrd 2007: 124). The second stage seeks answers to the question, "What is to be done?" If framed appropriately, these offer specific remedies and/or solutions to the problems raised in the first stage and the means for achieving them (Snow and Byrd 2007: 126). The final stage may be the most important. It seeks to motivate people to act (Snow and Benford 1988: 202). It seeks to overcome the fear of risks often associated with collective action and the free-rider problem (i.e., why participate in a risky and/or costly movement when I know someone else will, and that I will still benefit from their activism?) (Olson 1965; Snow and Byrd 2007: 128).

---

[3] The combination of these factors does not guarantee that a social movement will emerge, however.

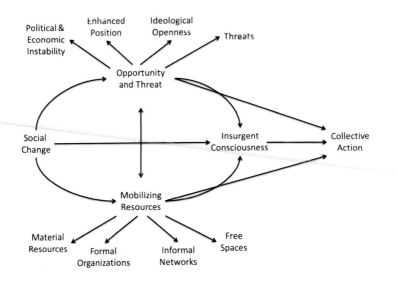

Fig. 1 Contemporary collective action model

Of course, a budding social movement's leaders generally do not expect that all potential members will be able to fully grasp the group's ideology or possess a sophisticated understanding of its history, which is why they often reduce their group's core message into generalized ideological snippets, bumper-sticker versions of the movement's broader ideology (Robinson 2004: 129), that are easily communicated to (and resonate with) potential followers and the general public (Snow et al. 1986). What becomes important, then, for those studying social movements is not only identifying the grievances that helped give rise to them but also how their leaders framed the movement's ideology in ways that helped them attract, retain, and motivate followers.[4]

Once frames are established, they can spread from one movement to another. Activists will often borrow slogans, songs, and various tactics from other movements. Diffusion can result with or without direct personal contact. When there is no direct contact, diffusion can occur when information is available through other channels, and the movements must have some elements in common (McAdam and Rucht 1993).

Scholars have noted that the media often plays an important role in interpreting social movement activism. Individuals, who are not directly involved in a particular protest or are geographically removed from it, often rely on some sort of media outlet, whether television, print, radio, or an Internet news source, to learn about it. Indeed, the news media often serve as a gatekeeper of movement information

---

[4] Given the 140 character limitation of Twitter messages (see discussion below), Twitter appears to be an ideal tool for the broadcasting of ideological snippets and the framing of grievances to a budding social movement's target audience.

(Koopmans 2004), choosing which events to report, which ones to ignore, and which ones to devote less time to. Interestingly, the news media are also somewhat dependent on social movements because the latter provide the drama, conflict, and action that sell newspapers and attract viewers. Nevertheless, evidence suggests that the media holds most of the power in the relationship (Gamson and Wolfsfeld 1993).

With mobile devices now able to access the Internet and broadcast to the entire world, the power relationship and the gatekeeper status appears to be changing. Twitter, a microblogging service that began in 2006, enables users to describe what they are doing, send a message to another user, embed a URL to another site, and broadcast their message to anyone else with a connection to the Internet. On its website, Twitter describes itself as a "real-time information network that connects you to the latest information about what you find interesting."[5] Anyone with a phone that has access to the Internet or SMS messaging (a form of text messaging) can send "tweets" from their phone, while those with computers with Internet access can post tweets through Twitter's website.[6] In addition to broadcasting tweets to others, users can choose to follow other users, so that they are constantly updated with new information when those they follow "tweet." Users can also describe their tweet as a topic, by placing a "#" before a word. If they want to pass on another person's message, they can "retweet" that message to their own followers or another user. Retweets contain the original message along with the original tweeter's name, but they can be passed on to other users with additional new content if the retweeting user so desires.

Kwak et al. (2010) studied the basic topology of Twitter. They found that a majority of the trending topics were headline or persistent news[7] and that retweeted messages reach, on average, 1,000 users, suggesting a fast diffusion of information across Twitter. Other research has found that although Twitter allows users to follow and communicate with many different people, users can really only entertain a maximum of 100–200 stable social relationships (Gonçalves et al. 2011), most of which reflect the process of homophily (i.e., "birds of a feather flock together"). For example, one study measured each user's general happiness and found that Twitter users exhibiting the same level of general happiness tended to associate with one another (Bollen et al. 2011). However, a study of political polarization on Twitter found that while highly polarizing messages that were retweeted tended to be sent to users sharing similar political leanings, messages sent directly to different users did cross political lines (Conover et al. 2011).[8]

---

[5] See http://www.twitter.com/about.

[6] A "tweet" is a message that is limited to 140 characters or less.

[7] Twitter tracks phrases, words, and phrases in order to find topics that become popular quickly and label as trending topics; these are published on its website and platform.

[8] Twitter's real-time capability has helped it be used in response to natural disasters both in the United States (Sutton et al. 2008) and abroad (Acar and Muraki 2011), response to acts of terrorism (Stelter and Cohen 2008), and for reactions to political debates (Shamma et al. 2009).

Social media also appears to be playing a larger part in protests although how, exactly, is unclear. For example, a 2010 report by the United States Institute of Peace noted that while social media is increasingly used as a tool for promoting freedom, there is still little understanding of its causal effects on political struggle (Aday et al. 2010). Social media can change the dynamics of mobilization by increasing speed and interactivity not found in traditional mobilization techniques (Eltantawy and Wiest 2011). Using the 2009 Iranian election protests as an illustrative case study, the authors show that social media not only played a role in the mobilization of protestors but also in the regime's own countermeasures (Aday et al. 2010).[9] As Evgeny Morozov (2011) notes, technology is a "double-edged" sword. It can be used just as easily to counter protest movements as it is to foster them.[10] Indeed, regimes from China to Egypt all employ technological barriers to communicate to one extent or another. Moreover, technology firms, including Twitter and even Google, have often failed to promote free political speech and instead struck deals with political authorities in order to facilitate business. This, however, has not kept activists from viewing social networking sites as a useful tool for collective action (Harlow and Johnson 2011).

Through continued pressure on Internet-based communications, authoritarian regimes are seeking to eventually attain something similar to their current control of traditional media (namely newspapers, radio, and television). In the past, oppressive regimes relied on the monopoly of information held by state-controlled traditional media to frame the discourse surrounding political action. One of the reasons why the aforementioned Kefaya movement failed was that state-controlled media portrayed the movement's goals as being daydreams funded by the U. S. Central Intelligence Agency (Oweidat et al. 2008). During the power of authoritarian regimes to control, information has been changing due to advances in information technology and the Internet. Minority groups or opposition groups that previously found it difficult to find information about their concerns now can find it online or delivered to their phone (Howard 2010).

Traditional media tends to favor the *status quo*, which in the case of the Arab Spring is the regime and leaders themselves, and focus on the negative aspects of protest movements. This tendency, sometimes referred to as the *protest paradigm* (Chan and Lee 1984) has been observed even in societies that are more liberal. McLeod and Detenber (1999: 3) analysis of traditional media coverage of protestors found that

> news stories about protest tend to focus on the protesters' appearances rather than their issues, emphasize their violent actions rather than their social criticism, pit them against the police rather than their chosen targets, and downplay their effectiveness.

---

[9] After officials declared incumbent Mahmoud Ahmadinejad the winner, supporters of opposition candidate Mir-Hussein Mousavi took to the streets to protest what they saw as a rigged election.

[10] See http://www.newscientist.com/article/mg20928026.100-the-internet-is-a-tyrants-friend.html.

This phenomenon is even more exaggerated in states with extensive control over the media. For example, after the Egyptian state-controlled media stopped ignoring the protests, it portrayed the protestors as agents of foreign meddling, radical Islam, Israel, or the West (International Crisis Group 2011). Repeatedly portraying protestors as misguided, delinquent, or violent, the media can (but not necessarily) delegitimize their grievances in the eyes of the public. The association of such negative qualities with protest movements can affect whether individuals join a movement or not. For example, an overrepresentation of negative impressions can increase the likelihood for individuals to adopt a similarly negative opinion of a movement (Iyengar 1990). However, it may not be feasible for governments to exert the same level of control over new media. Recent empirical work has shown traditional and social media to be functionally different as well. Harlow and Johnson (2011) show the content of Arab Spring-related social media posts was more positive and diverse than traditional media outlets, which tended to rely on government accounts.

Nevertheless, Twitter has helped alter the relationship between traditional news media and protesters. Protesters use Twitter to air their grievances and advertise their actions, and the news media use it to collect and verify information about newsworthy events. Hermida (2010: 302) suggests that journalists can use Twitter as an awareness system for journalists, allowing them to be alerted to "trends or issues hovering under the news radar." One case study looked at tweets by Andy Carvin, a journalist for National Public Radio, during the Egyptian Revolution and found that alternative voices had a greater influence over the content of Andy Carvin's tweets than other journalists or elite sources (Hermida et al. 2012). However, more "elite" traditional news organizations appear to be more reluctant to incorporate a new relationship between the news media and their audience than are less "elite" organizations (Lasorsa et al. 2011). Another case study that analyzed the content of tweets during the Egyptian Revolution found that Twitter can crowdsource the creation of information elites either by retweeting particular user's tweets or by encouraging other users to follow a particular user (Papacharissi and de Fatima Oliveira 2012). While traditional journalists became one trusted source of information, a parallel source of information was coming from accounts consisting of bloggers, activists, and intellectuals. This parallel source of information interacted with the traditional news media through tweets and retweets but also incorporated their opinions and emotions of solidarity with the protestors into the content of their tweets.

Is there a structural difference between traditional and social media that immunizes the latter to the *protest paradigm*? If such a mechanism were present, what form would it take? Below we offer one possible way to test for the effects of decentralized communication on this essential step of the political contention process.

4   Data, Methods, and Analysis

In order to understand the relationships among users of Twitter, we examined the pattern of tweets about Egypt during the Egyptian Revolution using social network analysis. Social network analysis is a useful method for researching discussion on Twitter because it can analyze the networks that emerge through response and retweeting patterns (Bruns and Burgess 2012). For example, centrality measures such as betweenness can help us identify users who are potential conduits of information, community detection algorithms can help us locate distinct clusters within the network, and triadic analysis can help us detect the presence (or absence) of complete triads (defined below) within such clusters. In this chapter, we tested whether particular aspects of clusters (i.e., which actors or what types of actors are most central) help explain structural features of those clusters (i.e., the number of complete triads). To be more precise, we tested whether activists who are in a position to broker the flow of information through a cluster are positively associated with greater cohesion within that cluster. If this proves to be the case, we argue that such results are consistent with the notion that highly central activists can use social media tools such as Twitter to frame grievances in ways that bring other actors together and mobilize them for action.

To test this, we analyzed over one million tweets about Egypt from January 28, 2011, and February 4, 2011, that were downloaded using the Twitter application programming interface (API), which is a function that allows other programs to receive and send data from and to Twitter. In our case, a program interacted with the Twitter API by requesting and storing all the data it could in regards to users tweeting about Egypt. Key aspects of the data that were collected and stored included the user's Twitter name, the content of each tweet, and a description of each user (from the user's public profile). Using these data, we generated a user-by-user network where a directed tie exists between two users if (1) one of the users sent a message to the other or (2) a user retweeted the message of another. In the case of the latter, a tie was drawn from the author of the original message to the user who "retweeted" the message. In the end, our user-by-user network included 196,670 users with 526,976 ties between them. From this, we extracted the largest weakly connected component that contained 176,447 users with 468,243 ties between them.

While the network as a whole is interesting, it is unlikely that an individual's framing of the events will transfer across such a wide network. Instead, the influence of a user will most likely have a larger effect among smaller clusters of users inside the larger conversation about Egypt. To identify and extract these clusters, we used a community detection algorithm developed by Blondel et al. (2008). This algorithm assigns users into different clusters in terms of the partition that yields the highest modularity score, which is a measure of fit developed by Mark Newman (2006) that compares ties within and across clusters to what one would expect in a random graph of the same size and having the same number of ties. Formally, it is the fraction of internal ties in each cluster less than the expected fraction, if they were distributed at random across the network. The higher the net fraction, the better the fit. The algorithm identified a number of distinct clusters. Clusters that contained at least 1,765 accounts or 1 % of the entire network were analyzed further. In this case, 30 communities contained at least 1,765 accounts.

We began by estimating betweenness centrality (Freeman 1979) of each of the users inside their individual communities. Betweenness centrality measures the extent to which each actor in a network lies on the shortest paths (i.e., geodesics) connecting all pairs of actors in the network. It is often used as a measure of brokerage; here we use it to identify users who are potential conduits of information and have the best opportunity to frame how information is portrayed. We examined the profiles of the top ten users and coded them as either traditional media (affiliated with television, radio, or print media), new media (affiliated as a blogger, online news, or portrayed himself or herself as a news source but lacking affiliation with traditional media), activist (self-identified as an activist or affiliated with human rights, social justice, political movement, or other movement), or other (e.g., celebrities, government accounts such as the State Department, users not portraying themselves as either activist or media, etc.). Table 1 summarizes the results of this analysis.

Next, we ran a triadic census for each community in order to examine their triadic structure.[11] A triad consists of three actors that may or may not be tied to one another.[12] Analyses of triadic dynamics generally focus on the property of *transitivity*, which describes the tendency for open triads to close, and *reciprocity*, which is when a tie exists in both directions. A triad is considered *transitive* if there is a directed connection A → B, B → C, and A → C, for any arrangement of the actors A, B, and C. Conversely, if the directed ties A → B and B → C both exist, but there is no A → C tie, the triad is considered *intransitive*.[13] Fritz Heider (1977) based his structural balance theory, the earliest study of transitivity, on the assumption that people seek to avoid cognitive dissonance, leading people with similar attributes or relationships to cluster together. This effect, known formally as *homophily*, is often described as "birds of feather flock together" or "the enemy of my enemy is my friend." Mark Granovetter (1973, 1983) has argued that transitivity occurs when strong ties are present among actors and tends to create small, tightly knit groups. *Reciprocity*, on the other hand, occurs when A has a tie to B, and B has a tie to A. *Reciprocity* is important because it is a source of social cohesion (Emerson 1962, 1972a, b, 1976). For the Twitter network we examine here, we consider a tie reciprocal if A sends a message to B and B sends a message to A and define a complete triad as one that is both transitive and all of its ties are

---

[11] The triadic census was estimated using the social network analysis program, Pajek (Batagelj and Mrvar 2012; de Nooy et al. 2005, 2011).

[12] Georg Simmel (1950) famously asserted that the triad is smallest irreducible unit of sociological analysis.

[13] For a full description of all possible triad types, see de Nooy et al. (2005: 209).

Table 1 Clusters from overall network

Cluster	Number of users	Average degree	Activists	New media	Traditional media	Other	Top user	Complete triads
1	19,062	3.422	7	3	0	0	New Media	282
2	13,645	1.641	1	2	7	0	New Media	84
3	11,421	1.250	1	1	5	3	Traditional Media	13
4	8,460	1.701	3	3	2	2	Other	2
5	8,069	1.333	0	0	6	4	Traditional Media	7
6	7,776	1.157	1	2	6	2	Traditional Media	0
7	6,946	1.474	4	6	0	0	New Media	4
8	7,375	1.218	2	1	6	1	Traditional Media	3
9	7,567	1.079	0	0	6	4	Traditional Media	0
10	6,036	1.212	5	3	0	2	New Media	0
11	5,875	1.291	1	2	6	1	Traditional Media	3
12	4,321	1.519	4	3	0	3	Activist	46
13	4,025	1.505	1	5	1	3	Other	2
14	4,216	1.342	1	3	2	4	Other	2
15	4,267	1.014	0	0	0	10	Other	0
16	3,754	1.380	0	2	8	0	Traditional Media	3
17	2,878	1.303	1	2	1	7	Traditional Media	3
18	2,969	1.171	0	4	5	1	Traditional Media	1
19	2,695	1.341	0	0	9	1	Traditional Media	14
20	2,280	2.318	7	3	0	0	New Media	20
21	2,329	1.284	0	2	4	4	Traditional Media	1
22	2,681	1.199	1	4	3	2	Activist	5
23	2,195	1.200	2	0	6	4	Traditional Media	3
24	2,090	1.282	2	0	3	5	Other	0
25	2,149	1.102	1	7	2	0	Traditional Media	0
26	1,899	1.450	2	1	3	4	Traditional Media	2
27	1,800	1.511	3	2	1	4	Activist	1
28	2,015	1.014	1	0	0	9	Other	0
29	1,875	1.126	1	5	1	3	New Media	1
30	1,768	1.278	1	0	6	3	Traditional Media	7

reciprocal. We focus on the presence of complete triads in this chapter because they are indicators of cohesion, and research has shown that members of cohesive subgroups tend to perceive (and by extension frame) similarly (de Nooy et al. 2005: 61; Everton 2012: 170). Interestingly (but perhaps not surprisingly), complete triads were the least common type of triad found in our analysis and were completely absent in some of the communities (see Table 1).

To determine whether the presence of activists, traditional media, or new media contributed to the presence of more complete triads, we estimated a multivariate count model that controlled for each community's number of users and average degree. We estimated a count model because the dependent variable is the count of the number of complete triads within each community. We used a negative binomial model because it best fit the distribution of counts within the dataset.[14]

Table 2 presents the results in terms of the incidence-rate ratios (IRR), which are similar to odds ratios in that they are equal to the natural log of the estimated coefficients (Long 1997; Long and Freese 2006). As the table indicates, when compared to number of "other" users (the reference category), the number of activists, members of the traditional media, and members of the new media ranked in the top 10 in terms of betweenness, centrality does not exert a statistically significant effect on the count of complete triads within a community. However, when the top user in each community is a member of the traditional media or is an activist, there is a positive and statistically significant effect on the count of complete triads in the community. In particular, when a member of the traditional media ranks first in terms of betweenness centrality, it increases the expected number of complete triads by more than three times than when an "other" user (the reference category) ranks first. A substantially larger effect occurs when activists rank first. When they do, it increases the expected number of complete triads by more than 11 times than when an "other" user ranks first. These results indicate that when activists or members of the traditional media are in positions of brokerage, the level of cohesion within a community is substantially greater than it would be otherwise. That highly central activists have the largest effect suggests that their position within the cluster may be providing them with the opportunity to use Twitter to frame events in ways that bring people together and possibly mobilize them. It is interesting that when a member of the new media ranks first, there is not a statistically significant increase in the number of complete triads. This may be due to some new media users being better at engaging other users than other new media users. However, the size of the new media IRR is almost as large as that of the traditional media, and its lack of statistical significance may simply be a function of the small sample size we are working with here. Moreover, since this is not a random sample, measures of statistical significance should be taken with a grain of salt (McCloskey 1995; Ziliak and McCloskey 2008). Indeed, because we

---

[14] We estimated the model using the generalized linear model (GLM) function implemented in StataCorp (2011), using a command by Hardin and Hilbe (2012), modifying the α so that the dispersion statistic was as close to 1.0 as possible.

Table 2 Multivariate regression results comparing the type of central accounts to the count of complete triads

	Model 1	Model 2
	IRR	IRR
Number of new media accounts in top 10	1.051	
Number of activist accounts in top 10	1.562	
Number of traditional media accounts in top 10	1.294	
New media as top account in cluster		3.245
Activist as top account in cluster		11.730*
Traditional media as top account in cluster		3.355*
Number of accounts in cluster	1.000	1.000*
Average degree in cluster	3.378	6.281*
AIC	6.010	5.701
BIC	−60.907	−54.166
Alpha dispersion	2.305	1.195

*$p < 0.05$

only analyze 30 clusters here, our conclusions should be considered tentative at best. Clearly, analyses of much larger Twitter datasets are in order.

5   Conclusions

In this paper, we have examined how social movements may be able to use Twitter to frame grievances in ways that resonate with their target audience. We saw how during the Egyptian Revolution, many people used Twitter to discuss the events that were taking place. We examined counts of complete triads, what we regard as an indicator of cohesion, and found them to be positively associated with the presence of activists or members of the traditional media in positions of brokerage. Because highly central activists exhibited the largest effect, we tentatively concluded that their network position appears to have helped them frame grievances in ways that brought people together and possibly helped mobilize them for action. Engaging other users and creating cohesive pockets of actors who view issues similarly increases the likelihood for a specific topic to become a trending topic and gain much more visibility.

Social movements will continue to use new tools that enable them to access a wider audience. In this case, people worldwide used Twitter to discuss events taking place in Egypt in early 2011. Not all social movements that have used Twitter have been successful. For example, Occupy Wall Street, a U.S. social movement, seeking to change the relationship between banks and the government, has yet to be successful although its members use Twitter frequently. While we do not claim that using Twitter will make a movement successful, it will be interesting to see whether other social movements that do achieve triadic closure are successful, and this closure is positively associated with the activists being in positions of information brokerage.

References

Acar A, Muraki Y (2011) Twitter for crisis communication: lessons learned from Japan's tsunami disaster. Int J Web Based Communities 7(3):392–402

Aday S, Farrell H, Lynch M, Sides J, Kelly J, Zuckermand E (2010) Blogs and bullets: new media in contentious politics. United States Institute of Peace, Washington, DC

Batagelj V, Mrvar A (2012) Pajek 2.04. University of Ljubljana, Lubjlijana

Blondel VD, Guillaume J-L, Lambiotte R, Lefebvre E (2008) Fast unfolding of communities in large networks. J Stat Mech arXiv:0803.0476v2

Bollen J, Gonçalves B, Ruan G, Mao H (2011) Happiness is assortative in online social networks. Artif Life 17(3):237–251

Bruns A, Burgess J (2012) Researching news discussion on twitter. Journal Stud:1–14

Chan JM, Lee CC (1984) The journalistic paradigm on civil protests: a case study of Hong Kong. In: Arno A, Dissanayake W (eds) The news media in national and international conflict. Westview Press, Boulder, CO

Conover MD, Ratkiewicz J, Francisco M, Gonçalves B, Flammini A, Menczer F (2011) Political polarization on twitter. In: Proceedings of the 5th international conference on weblogs and social media, Barcelona, Spain

de Nooy W, Mrvar A, Batagelj V (2005) Exploratory social network analysis with Pajek. Cambridge University Press, Cambridge

de Nooy W, Mrvar A, Batagelj V (2011) Exploratory social network analysis with Pajek. Revised and expanded. Cambridge University Press, Cambridge

Eltantawy N, Wiest JB (2011) Social media in the Egyptian revolution: reconsidering resource mobilization theory. Int J Commun 5:1207–1224

Emerson RM (1962) Power-dependence relations. Am Sociol Rev 27(1):31–41

Emerson RM (1972a) Exchange theory, part I: a psychological basis for social exchange. In: Berger J, Zelditch M, Anderson B (eds) Sociological theories in progress. Houghton-Mifflin, Boston, MA, pp 38–57

Emerson RM (1972b) Exchange theory, part II: exchange relations and network structures. In: Berger J, Zelditch M, Anderson B (eds) Sociological theories in progress. Houghton-Mifflin, Boston, MA, pp 58–87

Emerson RM (1976) Social exchange theory. In: Annual review of sociology. Annual Reviews Inc., Palo Alto, CA, pp 335–362

Everton SF (2012) Disrupting dark networks. Cambridge University Press, Cambridge and New York, NY

Freeman LC (1979) Centrality in social networks I: conceptual clarification. Soc Networks 1:215–239

Gamson WA, Wolfsfeld G (1993) Movements and media as interacting systems. Ann Am Acad Pol Soc Sci 528:114–125

Gonçalves B, Perra N, Vespignani A (2011) Modeling users' activity on twitter networks: validation of Dunbar's number. PLoS ONE 6(8):e22656

Granovetter M (1973) The strength of weak ties. Am J Sociol 73(6):1360–1380

Granovetter M (1983) The strength of weak ties: a network theory revisited. Sociol Theory 1:201–233

Hardin JW, Hilbe JM (2012) Generalized linear models and extensions, 3rd edn. Taylor & Francis, London

Harlow S, Johnson TJ (2011) Overthrowing the protest paradigm? How The New York Times: global voices and twitter covered the Egyptian revolution. Int J Commun 5:1359–1374

Heider F (1977) Attitudes and cognitive organization. J Psychol 21:107–112

Hermida A (2010) Twittering the news. Journal Pract 4(3):297–308

Hermida A, Lewis SC, Zamith R (2012) Sourcing the Arab Spring: a case study of Andy Carvin's sources during the Tunisian and Egyptian revolutions. Paper presented at the international symposium on online journalism, Austin, TX

Howard PN (2010) The digital origins of dictatorship and democracy: information technology and political Islam. Oxford University Press, New York, NY

Howard PN, Hussain MM (2011) The upheavals in Egypt and Tunisia: the role of digital media. J Democracy 22(3):35–48. Retrieved from http://www.journalofdemocracy.org/upheavals-egypt-and-tunisia-role-digital-media

International Crisis Group (2011) Popular protest in North Africa and The Middle East (I): Egypt victorious? International Crisis Group, Brussels

Iyengar S (1990) The accessibility bias in politics: television news and public opinion. Int J Public Opin Res 2(1):1–15

Koopmans R (2004) Movements and media: selection processes and evolutionary dynamics in the public sphere. Theor Soc 33:367–391

Kwak H, Lee CC, Park H, Moon S (2010) What is twitter, a social network or a news media? In: Proceedings of WWW 2010, Raleigh, NC

Lasorsa DL, Lewis SC, Holton AE (2011) Normalizing twitter. Journal Stud 13(1):19–36

Lim M (2012) Clicks, cabs, and coffee houses: social media and oppositional movements in Egypt, 2004–2011. J Commun:231–248. http://onlinelibrary.wiley.com/doi. doi:10.1111/j.1460-2466.2012.01628.x/abstract

Long JS (1997) Regression models for categorical and limited dependent variables, Advanced Quantitative Techniques in the Social Sciences Series. Sage Publications, Thousand Oaks, CA

Long JS, Freese J (2006) Regression models for categorical dependent variables using stata, 2nd edn. Stata Press, College Station, TX

Lynch M (2011) After Egypt: the limits and promise of online challenges to the authoritarian Arab state. Perspect Polit 9:301–310

McAdam D (1982) Political process and the development of black insurgency, 1930–1970. University of Chicago Press, Chicago

McAdam D, Rucht D (1993) The cross national diffusion of movement ideas. Ann Am Acad Polit Soc Sci 528(1):56–74

McAdam D, Snow DA (eds) (2010) Readings on social movements: origins, dynamics, and outcomes. Oxford University Press, New York, NY and Oxford

McAdam D, McCarthy JD, Zald MN (1988) Social movements. In: Smelser NJ (ed) The handbook of sociology. Sage, Beverly Hills, CA, pp 695–737

McAdam D, Tarrow S, Tilly C (2001) Dynamics of contention. Cambridge University Press, New York, NY and Cambridge

McCarthy JD, Zald MN (1977) Resource mobilization and social movements: a partial theory. Am J Sociol 82(6):1212–1241

McCloskey D (1995) The insignificance of statistical significance. Sci Am April:32–33

McLeod D, Detenber B (1999) Framing effects of television news coverage of social protest. J Commun 49:3–23

Morozov E (2011) The Internet is a Tyrant's Friend. New Sci. Retrieved from http://www.newscientist.com/article/mg20928026.100-the-internet-is-a-tyrants-friend.html

Newman MEJ (2006) Modularity and community structure in networks. Proc Natl Acad Sci 103 (23):8577–8582

Olson M (1965) The logic of collective action: public goods and the theory of groups. Harvard University Press, Cambridge, MA

Oweidat N, Benard C, Stahl D, Kildani W, O'Connell E, Grant AK (2008) The Kefaya movement: a case study of a grassroots reform initiative. RAND Corporation, Santa Monica, CA

Papacharissi Z, de Fatima Oliveira M (2012) Affective news and networked publics: the rhythms of news storytelling on #Egypt. J Commun 62(2):266–282

Robinson GE (2004) Hamas as social movement. In: Quintan W (ed) Islamic activism: a social movement theory approach. Indiana University Press, Bloomington, IN, pp 112–139

Salama A (2011) How to advertise a revolution. TEDxCairo. Retrieved from http://www.tedxcairo.com/index.php?show=Talks&id=6

Shamma D, Kennedy L, Churchill EF (2009) Tweet the debates: understanding community annotation of uncollected source. In: Proceedings of WSM'09, Beijing, China

Simmel G (1950) The triad. In: Wolf KH (ed) The sociology of Georg Simmel. The Free Press, New York, NY, pp 145–169

Smith CS (1991) The emergence of liberation theology: radical religion and social movement theory. University of Chicago Press, Chicago

Snow DA, Benford RD (1988) Ideology, frame resonance, and participant mobilization. Int Soc Mov Res 1:197–217

Snow DA, Byrd SC (2007) Ideology, framing processes, and Islamic terrorist networks. Mobilization 12(1):119–136

Snow DA, Burke Rochford E Jr, Worden SK, Benford RD (1986) Frame alignment processes, micromobilization, and movement participation. Am Sociol Rev 51(4):464–481

StataCorp (2011) Stata statistical software: release 12. StataCorp LP, College Station, TX

Stelter B, Cohen N (2008) Citizen journalists provided glimpses of mumbai attacks. N Y Times. Retrieved from http://www.nytimes.com/2008/11/30/world/asia/30twitter.html

Sutton J, Palen L, Shklovski I (2008) Back-channels on the front lines: emerging uses of social media in the 2007 Southern California wildfires. In: ISCRAM conference, Washington, DC

Ziliak ST, McCloskey DN (2008) The cult of statistical significance. The University of Michigan Press, Ann Arbor, MI

The Arab Spring in North Africa: Still Winter in Morocco?

Rebecca S. Robinson and Mary Jane C. Parmentier

Abstract This chapter focuses on Morocco, presenting research conducted on the Moroccan blogosphere between 2007 and 2009, before the Arab Spring, in which a burgeoning civil society space with some evidence of collective action was apparent. Despite this online activism and the highest level of Internet penetration in North Africa, the Moroccan government withstood Arab Spring protests. While the religious and political legitimacy of the Moroccan monarchy is a key factor in explaining this difference, there are also significant socioeconomic and linguistic cleavages that were mirrored in the online space. Nevertheless, the Moroccan blogosphere leading up to the events of 2011 provides a rich public narrative that includes collective action unique to this historically authoritarian North African country. The chapter will utilize selected collective action literature to examine the nature of this narrative and then examine it in light of several Arab Spring protests that did occur in Morocco, concluding that, while the monarchy retains its power and online collective action does not reach the majority of citizens, online social and political expression do represent a potentially significant development in Morocco's public sphere.

1 Introduction

As the governments of Tunisia, Egypt, and Libya fell in 2011, political protests swept the entire Maghreb region. Common themes were apparent, such as distress over the lack of economic opportunities, disgust with government corruption, and mobilization of activists and participants in a growing online social media

R.S. Robinson (✉) • M.J.C. Parmentier
Arizona State University, Tempe, AZ 85287, USA
e-mail: rebecca.robinson@asu.edu

N. Agarwal et al. (eds.), *Online Collective Action*, Lecture Notes in Social Networks,
DOI 10.1007/978-3-7091-1340-0_11, © Springer-Verlag Wien 2014

environment. And yet, the governments of Morocco and Algeria remain, while those of Tunisia, Egypt, and Libya have undergone rapid change. There are differences in political culture and government legitimacy that help account for this apparent anomaly, but these countries in which changes were less dramatic offer interesting cases in which to study online collective action. One goal of this chapter is to explain why the collective action frames of the Moroccan protests before and during the Arab Spring that were disseminated through the Francophone Moroccan blogosphere, the Blogoma, did not lead to significant threats to the monarchy.

Morocco provides an apt example of how the digital divide prevents researchers of online collective action from understanding the larger society, which is socio-economically, ethnically, and linguistically fragmented, because the digital divide encompasses splits endogenous and exogenous to the online universe that mirror already extant splintering within the society. Literature on the digital divide and collective action frames will be discussed to elucidate why the ongoing collective action of the Blogoma, which primarily focuses on civil liberties and free speech, did not appeal to the lower class, majority of Moroccans during the Arab Spring. The online and offline collective action frames of the Arab Spring in Morocco, as they were articulated through the Blogoma, did not provide a holistic picture of the sentiments of the larger Moroccan society because these frames were of import to a well-mobilized but relatively small and more affluent sector of the population. Unlike Tunisia, where research has shown that, despite a socioeconomically fragmented society, the fiery death of a fruit vender was effectively "framed" as an injustice that resonated with the wider community (Lim 2013), in Morocco, issue of free speech on the Internet and other civil liberties did not seem to connect with the wider society.

2   Digital Divide and Collective Action Frames

Selwyn (2004) argues that the digital divide is not dichotomous but rather is characterized by how individuals access and make use of information and communications technologies (ICTs), influenced by their cultural and social experience and context. While Morocco is often perceived as a relatively cohesive, homogeneous society, there are important socioeconomic, political, ethnic, linguistic, and religious distinctions, despite the fact that nearly all Moroccans would identify themselves as such and as Muslims. The literature on the digital divide and collective action seems to overwhelmingly take issue with the notion of a dichotomous, bi-polar divide between those with access to the Internet and those without. Warschauer (2004) has pointed out that Internet diffusion and use takes place within particular socioeconomic, cultural, and political contexts; however, it is not a "bipolar societal split" (p. 7). It is argued that the reality is much more complex with collectivities online replicating or being influenced by social, cultural, and socioeconomic divides.

Graham (2011) portrays a complex and nuanced cyberspace, taking into account "economic, cultural, political and technological" (p. 212) aspects of each person accessing it. One telling example of this complexity in the Moroccan context is the socioeconomic divide that is also linguistic, in that the middle and upper classes are educated in French and prefer to write in French rather than Arabic. So even lower class Moroccans with Internet access may not have sufficient proficiency in French (or even written Arabic) to gain entry into the Blogoma. Thus, in the Moroccan context, the digital divide does not only exist between online and offline communities but also within networked environments.

There was evidence in the 2007–2009 research that some Arabophone bloggers had different concerns from their Francophone counterparts, such as BlaFrancia's blog (transliterated from Arabic as "Without French"), in which the main purpose is to encourage Moroccans to use Arabic (substantiated by Etling et al., 2009). Arabic bloggers were less inclined than their Francophone counterparts to discuss the Mourtada and Erraji incidences (discussed in detail below) and may be more religiously and/or politically conservative, according to some of the respondents of the second survey in 2009. It is possible that Arabophone bloggers are more attuned to the sentiments of the larger population, but more research is necessary to confirm this possibility.

Other scholars suggest that in addition to online discourse mirroring offline societal divisions on sociopolitical issues, these online schisms might also exacerbate the existing social and political divides. Farrell (2012), questioning the assertion that online collective action can reduce political divisions, suggests that it can actually increase polarization by encouraging "homophily," which involves only associating with those who have similar views and experiences (p. 41). Couldry reiterates this assertion, finding that certain groups build their own spaces online, as they do offline, which can replicate and reinforce existing political divisions (2003, p. 94). Therefore, the interactions within the Blogoma could serve to reaffirm within this community that the topics they discuss, censorship and freedom of speech, are the most important issues facing Moroccan society.

Considering the possibility that the Blogoma is a network characterized by homophily, online collective action may differ significantly from offline collective action because the former may present a unified front, while the larger society is much more fragmented. With this in mind, one should reflect on the distinctions between the ways that collective action was framed in the Blogoma during the Arab Spring and the actual outcomes of the protests, referendum, and elections that occurred in Morocco in 2011.

Two issues are of principal importance in discussing how the digital divide affects online and offline collective action frames: prognostic frames, which some scholars argue are intricately tied to the diagnostic frames (Gerhards and Rucht, 1992), and frame resonance. Benford and Snow define collective action frames as "emergent action-oriented sets of beliefs and meanings that inspire and legitimate social movement activities and campaigns" (2000, p. 614). First, prognostic frames are what the collective actors propose to solve the problems, which they identify through the diagnostic frames, often referred to as "injustice framing" (Gamson

1992, p. 68), which are what or who the collective actors identify as the cause(s) of the problem. Some of the diagnostic frames of the Blogoma, which also appeared to inform some of the collective action during the Arab Spring protests, were identified in the past research conducted between 2007 and 2009: the bloggers wanted to expand the Moroccan repertoire of civil liberties, including free speech, and install a government that would be accountable to the people.

As will be discussed in greater detail, some of the actors associated with the Blogoma and the protests during the Arab Spring called for the creation of a parliamentary monarchy—their prognostic frame (Des maux a dire, 2012). Clearly, the diagnostic frames of the arbitrary powers of the monarchy, the corruption and unaccountability of the government, and the lack of civil rights allotted to Moroccans were integral in formulating this prognostic frame. The diagnostic frames were also identified during the 2007–2009 research project.

A few of the critical factors influencing frame resonance are salience and credibility (Benford and Snow 2000, p. 619). While the Blogoma undoubtedly collectivizes around important concerns, these concerns may not be the main concerns of the majority of the Moroccan population. For example, since bloggers participate in the media landscape of the nation, they want to be able to discuss any issue without fear of reprisal, but they are not likely to be acutely affected by the endemic poverty, lack of social services, and the pervasive paucity of economic opportunities, even though a substantial portion of middle/upper class Moroccans are also unemployed. Therefore, the ways in which bloggers frame online collective action (OCA) could lack salience among the majority of the population.

The credibility of frames plays an important role in how the Arab Spring played out in Morocco because the makhzen (the elite power structure) used this upheaval to demarcate "authentic" Moroccans from the inauthentic ones. The official propaganda released during the protests made claims that foreign entities or Moroccans living abroad were seeking to destabilize the country. This propaganda called upon "real" Moroccans to uphold their civic duties through voting (rather than protesting) and to respect the sacred Moroccan institutions: the monarchy, Islam, and the stability of the state.

Although the Blogoma represents a mobilized portion of the population, it is perhaps too small and literally inaccessible, through the digital divide, to reach the majority of Moroccans. One must keep in mind that nearly half of the Moroccan population is illiterate and that an estimated 40–50 % of the Moroccan population is of Amazigh descent (Maddy-Weitzman, 2001), which signifies that, while some of the Amazigh men may communicate in Arabic, the women generally do not communicate in either French or Arabic. Thus, the majority of Moroccans most likely only have access to official propaganda and the collective action frames of officially recognized political organizations, such as the Party of Justice and Development (PJD), because, as will be discussed in the case of Erraji, the government keeps tighter reigns on social media and the independent press in Arabic. The credibility of the PJD has not been tainted by the makhzen's smear tactics, as was the case with the protesters, or by past involvement in unsavory dealings and corrupt practices like most of the well-established political parties in the Moroccan parliament. However, unlike the role of the Muslim Brotherhood in coalescing descent of the Mubarak regime in Egypt (Lynch, 2009), the preeminent Islamic party has largely been coopted by the government.

3   Methods

The initial research conducted by the lead author between 2007 and 2009 employed online ethnography, which included an inductive content analysis, interviews, and participant observation. Open-ended surveys were employed to include the opinions of a greater number of Blogoma interlocutors and to draw it the participation of the Arabophone bloggers, which were initially excluded from the content analysis due to the researcher's linguistic limitations.

Most qualitative research on online communities is limited to content analysis, which may or may not be substantiated through in-depth interviews. However, the preference for these two methods bypasses one important step—participant observation. Barriers associated with entrée into online communities can be mediated through establishing rapport with gatekeepers and by mitigating suspicion about the researcher's intentions through prolific participant observation, which can be accomplished in various ways: creating blogs, commenting regularly on the blogs of potential research subjects (and linking back to one's own blog to disseminate information about the research and the researcher's position on various issues), and developing profiles on various social networking sites.

The second method employed in the online ethnographical methodology of the Blogoma was content analysis. Through inductive content analysis, the researcher looks for events or topics that garner many responses on the most trafficked websites. Reputable sources in communities are highly linked to by other community members and create nodal points of interaction. Furthermore, in the case of Morocco, bloggers organize an annual ceremony, the "Maroc Blog Awards," so reputable sources can also be found among those recognized with awards. Due to the inductive nature of the content analysis, the researcher opted for a topically oriented analysis rather than an event-based one. Content analysis is crucial to asking interview questions that elaborate on topics discussed on the blogs and to clarify points of interest to the researcher.

The online ethnography employed in-depth interviews to elaborate on the content analysis; however, interviews are not possible without first establishing one's presence in the community through participant observation and understanding which topics are of interest to the community through the content analysis. In the case of Morocco, interviews were useful for determining what content was not included on the blogs through self-censorship and for building off of data from previous interviews. Thus, the combination of content analysis, participant observation, and in-depth interviews, or use of multiple methods toward triangulation, results in thorough, rigorous and textured data.

In addition to triangulation, another important aspect of enhancing the credibility of qualitative research is the length of time in the field. For this study of the Blogoma, the researcher was "hanging out" for over a year before the findings were analyzed. It was not until the writing phase of the analysis that the distinctiveness of the Francophone blogosphere from the Arabophone one became apparent—one of the most interesting findings of the research. The point is that in relatively unfamiliar spaces, the longevity of research is key to developing interesting and well-informed data.

It was fairly straightforward to update the research to include the events of 2011 because many of the major contributors of the Blogoma remained the same. The authors looked at the most popular French and English blogs included in the original research. These blogs were utilized to determine new actors who were active during the protests. The authors employed an event-based approach in this research because the topics of interest had already been identified through the content analysis: support for street protests, proposed reforms, diagnostic and prognostic frames, and general election. The blogs were examined between January and March 2011 (the onset of the Arab Spring), June and August 2011 (the voting on the referendum), and October and December 2011 (the general elections).

4 The Context: The Moroccan Blogosphere

The Blogoma came to life in 2004. In 2007, there were an estimated 25,000 blogs in Morocco (Bensalah, 2007) and about 2,000 Francophone blogs actively aggregated under "Maroc Blogs," which was the platform that was primarily consulted during the initial research. In 2009, the President of the Association of Moroccan Blogger, Said Benjebli, stated that there were 30,000 blogs in Arabic and slightly more in French (Belhaj, 2009). In 2010, Morocco had the highest rate of Internet use in North Africa, with 49 users per 100 people (ITU, 2011). This is somewhat surprising due to the fact that Morocco had the lowest Human Development Index (HDI) in North Africa in 2011, at 130.

King Mohamed VI has made serious commitments to building Morocco's technological capabilities. By 2003, the government had invested $1.2 billion (or $11.90 per capita) to develop and enhance the digital network (Ibahrine, 2004). In 2007, the World Bank's infoDEV program reported that Morocco has focused on ICTs in education since 1999, as a means of enhancing the country's global competitiveness. Ironically, the makhzen have been pushing for ICT development; however, it stands to lose the most by users of this expanded network collectivizing against it, which is certainly an important motivator of the formation of the Blogoma. The makhzen have opened Pandora's laptop.

5 Online Collective Action in the Blogoma

Although the Blogoma was initially founded as a social network (Myrtus, 2008), it has morphed into a vehicle of collective action through disseminating information, countering censorship, advocating for civil rights, political blogging, and mobilizing social initiatives. Though these initiatives have met varying degrees of success, the Blogoma has been most outspoken about issues of free speech and civil rights. This section draws on the research conducted between 2007 and 2009 to elucidate the ongoing campaigns of the Blogoma.

Even in its nascent stages, the Blogoma had become an influential player in the Moroccan media landscape. While the independent press must be extremely careful about what it discusses, the state appears to be much more lenient with bloggers. Some Moroccan bloggers have even reached a level of prominence suggestive of celebrity status. In 2007, Larbi, writing under his metonym, was cited as one of "Les 100 Qui Font Bouger le Maroc" ("100 People Who Move Morocco") (El Hilali, 2007). Mohamed Drissi Bakhkhat, previously blogging at *MoTIC*, which was the 2008 Maroc Blog Awards winner for the best "IT Blog," claims that traditional media have been affected by the Blogoma: "reaction throughout the Moroccan blogosphere [about censorship issues] inspired many articles in almost all of the Moroccan independent press" (Ben Gharbia, 2007). His statement draws attention not only to the Blogoma's influence on other media but also to the high degree of mobilization around censorship issues.

While the efficacy of the Blogoma's collective action toward promoting civil liberties is more ambiguous than its influence on the Moroccan press, it is through the civil rights campaigns that the Blogoma space has gained the attention of the international press and, along with its campaigns against censorship, mobilized the greatest number of individual actors. The bloggers rallied against the prosecution of engineer, Fouad Mourtada, for creating a fictitious Facebook profile under the name of the king's brother in January of 2008 (El Hilali, 2008a). Three weeks after the creation of the profile, Mourtada was disappeared, during which time he was interrogated and tortured (El Amrani, 2008). The Blogoma started discussing his arrest on February 6, the day after Mourtada was arrested, facing the charge of "villainous practices linked to the alleged theft" of the Crown Prince's identity, which carried a prison sentence of up to 5 years (Ben Gharbia, 2008). On February 19, the Blogoma initiated a strike to show solidarity with Mourtada; he was, nonetheless, sentenced to 3 years in prison and fined 10,000 dirham on February 22 (El Hilali, 2008b).

The Blogoma continued to protest after Mourtada's sentencing, which may have contributed to his royal pardon on March 18 (York, 2008). The jailing of this average citizen affected the Blogoma in a profound way because the incident spoke to the limitations of freedom that could be exercised in the more liberated (than most national blogospheres in the region) Moroccan blogosphere; however, some viewed the news of the pardon as a major victory. Some bloggers extolled the solidarity of the Blogoma as a force for progress. Others pointed out Mourtada's release was a display of royal power – it was a reminder that freedom is the king's prerogative (Eatbees, 2008). The arbitrary power of the king perhaps explains why some bloggers, like Bakhkhat, decided to stop blogging after this incident. Others claimed that it was not the influence of the Blogoma but international pressure that led to the pardon.

Another event in 2008 that caught the attention of the Blogoma was the arrest of Mohammad Erraji. Erraji wrote regularly for hepress.com, an online journal in Arabic. On September 3, he published a piece online entitled "The King Encourages the Dependency of his People" (Lalami, 2008). In the article, Erraji attacked the widespread practice of *grima*, which is the Moroccan practice of begging for official capacities and favors rather than earning them through work and merit. Laila Lalami noted that this topic would not be overly risky if broached in an independent, Francophone publication, but Erraji's publication in Arabic accessible to the lower echelons of society, the literate portion at least, was more than the existing power structure could handle (ibid). According to the account on Erraji's blog, he was arrested on September 5 (Erraji, 2008). Erraji's "trial" took place on September 8. Reporters Without Borders called the legal proceedings "worthy of the most totalitarian states," an assertion that was reiterated by other nongovernmental organizations as well (RSF, 2008). He was remanded to custody for 2 years and ordered to pay a fine of 5,000 dirham, which seems like a pittance when other Moroccan journalists have been fined hundreds of thousands of Euros but, to an average citizen, it would be enough to bankrupt him. This possibility is substantiated by Erraji's statement to the judge that he could not afford to hire a lawyer (Cherkaoui, 2008).

In this case as well, it was the Blogoma that started the impetus for mainstream and international media coverage of the event. The Blogoma had been discussing Erraji's detainment for days, while the mainstream press picked up the story only after he had been convicted. The bloggers organized a 24 hour strike on September 15, in which the lead author participated (York, 2008), started an online petition demanding his release (http://helperraji.com/), and created a group on Facebook (http://www.facebook.com/groups/30771925854/members/) (soutienerraji.blogspot.com, 2008). It is possible that these efforts contributed to his eventual acquittal in appeals court on September 18.

In contrast to these successful efforts related to expanding the purview of free speech and civil rights, the Blogoma has had little success influencing the political establishment. It is in this arena that the problematics of the digital divide become most apparent. There is little evidence that the political opinions of the Blogoma reach other Moroccans living inside of the country, especially those who predominantly read Arabic. While many of the bloggers either encouraged abstention or votes against the PJD in the 2007 parliamentary election, the outcomes stood in stark contrast to the desired results of the Blogoma—the voter turnout was quite good and the PJD ended up doubling its seats in the parliament.

Despite the Blogoma's poor record in influencing election outcomes, it has undertaken some successful social campaigns: some predominantly online, like SOS Morocco and *Blogons Utile*; and others that straddled both the online and offline environments. In 2009, the Blogoma in partnership with offline activists were able to amass two tons of medical supplies to send to the Gaza Strip (Lady Zee, 2009). However, while members of the Blogoma are anxious about certain social issues facing the Moroccan society, like the educational system, these issues do not directly affect their lives. This contributes to the evident disjuncture between the topics of interest in the Blogoma and the issues that affect the overwhelming majority of Moroccans. This is not to say that the bloggers are apathetic toward the poorer echelons of society but that their activism—centering on issues of free speech, civil rights, and, more recently, political reform—speaks to the possibility that the bloggers believe that the political structure and the control of information would have to be drastically altered before the establishment would become politically, socially, and materially accountable to the average Moroccan citizens.

While campaigns dedicated to securing civil liberties were most prevalent during the lead author's past research, several campaigns related to the censorship of various websites occurred in 2006 and 2007: pro-separatist websites and Google Earth were blocked in 2006, as were antimonarchical websites and YouTube in 2007. Nevertheless, this important topic was highlighted in the survey responses. Eighteen of the twenty-one respondents to the first survey considered incidents linked to the issues of free speech and censorship to be the most important events that have occurred in the Blogoma. Through implementing grounded theory, the lead author was able to devise a more appropriate survey the second time, which increased the number of respondents from 11 completed surveys in the first round (although more than 20 partial surveys were submitted) to 47 in the second. Forty-two respondents unanimously answered "Yes" to the following question: should people be able to debate about the three sacred institutions that are protected under the Moroccan constitution—Islam, the monarchy, and national integrity, which primarily focuses on the Moroccan/Western Sahara? This indicates that Moroccan bloggers want the freedom to address anything they wish, thereby establishing the ability to guard against abuses that occurred in the past under the former King Hassan II, who died in 1999, and ensuring their rights in the present and future.

Also related to the issue of free speech is the pervasiveness of self-censorship in the Blogoma: 26 of 43 respondents said that they are careful about what they write on their blogs. This tendency has been more rigorously observed since the arrests of Mourtada and Erraji; tellingly, 35% of respondents claim to have become more careful about what they say. It should be noted, however, that some Moroccan bloggers are writing outside of Morocco where there is little concern for retaliation against them. As a side note, the separation between the real and virtual worlds holds less weight in more oppressive regimes—apparent in the cases of Mourtada and Erraji; as such, self-censorship is a very real concern for researchers who conduct studies on populations in more tightly controlled political environments. Researchers must recognize that methods used to conduct research in the West cannot be facilely redirected to other nations without paying attention to the specific cultural contexts.

6   Influencing Offline Politics and Social Movements

Morocco experienced mass street protests on February 20, 2011, coined the February 20 Movement, which saw thousands of Moroccans, largely youth, in the streets, particularly in the capital of Rabat (Alami, 2012; Klein, 2013). The protesters demanded improved education and standards of living but also a political structure that is accountable to them. They wanted to be free from the prerogative of the king to imprison and liberate, such as in the Mourtada case, and from cronyism that functions through special favors and connections, described in Erraji's discussion of *grima*. The organization and culmination of the mass protests has been attributed to the leaders organizing and communicating on Facebook beginning in 2009 but also to the online activism of bloggers before 2011 (Rahman, 2012). The Blogoma, then, could be seen as a precursor to OCA on social media platforms such as Facebook and Twitter.

While some of the specific connections between the Blogoma and the February 20 Movement are difficult to identify, others are more apparent. Hisham Almiraat, for instance, who blogs at "Moroccan Mirrors" and is a contributor to Global Voices Online, joined with the February 20 youth movement to develop the Moroccan Pirate Party (Errazzouki, 2012). The primary goals of the party are education, transparency, and rule of law (ibid). Almiraat also cofounded the Mamfakinch website, available in French and Arabic, which announces, maps, and reports on street demonstrations. It is, according to the website, produced by Moroccan "bloggers and militants" and has attempted to offer independent coverage of the protests (Mamfakinch, n.d.). This is significant in a country in which the government is so consumed with promoting its positive international image that it has allegedly turned hundreds of thousands of protesters (300,000 by protest organizers' accounts) into tens of thousands (official calculations) through its stranglehold on the official press (M4C, 2011a). The stifled press is also clearly an important concern of this movement and Blogoma activists.

Since the regime refrained from oppressing the February 20 Movement at its onset for fear that doing so might create another Tunisian or Egyptian style revolt, it used smear tactics to present the protesters as agents of a foreign entity or the Western/Moroccan Saharan insurgency group (Polisario). It also framed them as traitorous, inauthentic Moroccans, such as "atheists … republican revolutionaries … [and] Moroccans living comfortably abroad" (Lalami, 2011a) perhaps alluding to certain expatriate bloggers. King Mohammad VI responded with a recipe for cosmetic constitutional reforms, doubling food subsidies, and employing 4,000 unemployed graduates in a country of over 30 million people (ibid). One commentator claimed that the government postponed voting on the new constitution for 3 months in an effort to wait out the momentum of unrest both domestically and abroad (Imad, 2011a). Although the regime initially allowed the peaceful demonstrations to occur without retribution, a month after the onset, when the protests failed to fizzle out, the government began to be less tolerant of the right of the people to assemble.

This discussion still begs the question as to why events in Morocco have not escalated to a tipping point as they did in Tunisia, Egypt, Libya, and elsewhere. Several issues become apparent. Firstly, Mohammad VI, or the monarchy in general, has maintained popularity with the population even though his lifestyle costs the Moroccan people about $270 million per year—more than the British monarchy (Lalami, 2011b). The Alaouais have reigned in the area of present-day Morocco since the seventeenth century, and their lineage to the Prophet Mohammad reifies their legitimacy. If these factors fail to impress, one blogger points out the well-rehearsed indoctrination with which Moroccans are inculcated from a very young age (Roumana, 2011)—the end result is the inseparability of the king from country and God.

Secondly, the king is seen by many as instrumental in solidifying an otherwise incongruent group of people into a nation. The bombing in Marrakech on April 28, 2011 that killed 17 people (M4C, 2011a) was a convenient reminder of the more radical elements of the Moroccan society, though few in number, that may, according to official discourse, seek to destabilize the country. According to one blogger, rulers throughout the Muslim world have used the supposed threat of Islamists infiltrating the power structure to squelch popular assembly (Hogan, 2011).

Thirdly, February 20 appears to function on a part-time basis. One commentator points out that one of the main differences between the movements in Morocco and those of Tunisia and Egypt is that people in the latter cases took to the streets and stayed there until they achieved their goals (Imad, 2011b). Moroccan activists have tended to set aside various Sundays to voice their grievances rather than maintaining a constant presence in the streets. The February 20 Movement has also faced internal leadership conflicts and failed to collaborate with other civic organizations that were demanding change. "Moroccans for Change" listed 37 different organizations that have taken shape since the onset of the Arab Spring, reminiscent of the fragmentation of the Moroccan parliament (M4C, 2011b).

Finally, aside from the evident splintering among political affiliations, with the tendency of the lower classes to be more accepting of Islamic-oriented political parties, some bloggers claim that the Moroccan society lacks social cohesion and the middle class in general have little faith in the ability of the lower classes to effectively participate in a democratic polity. Middle class constituents instead, according to Tourabi (2011), attempt to hold on to their nominal privilege through any means at their disposal, which essentially eliminates any possibility of them uniting with the lower socioeconomic classes.

Since the Blogoma and leaders of the February 20 Movement reject the integration of religion into politics, evident through their proposed boycott of the parliamentary elections of 2011 (Almiraat, 2011; El Hilali, 2011), the people that voted overwhelmingly in support of the PJD were most likely the lower class majority of Moroccans. Some of the bloggers stated that the PJD was the only party that has not been tainted with an unsavory history of corruption and cronyism, but they still promoted boycotting the November 25 elections rather than voting for PJD, which was forecasted to be the most successful contender in the election. It is possible that the supporters of the PJD believe that an Islamic-oriented party may act in a more responsible and charitable manner to the millions of lower class Moroccans, like Fadoua Laroui, who found the prospects of voicing her despair through self-immolation more attractive than living in a society that had dehumanized her. The PJD has promised to decrease poverty by 50%, increase minimum wage by 50%, and eradicate corruption (Cabalamuse, 2011), objectives which speak to concrete goals of improving the material conditions of people's lives.

Less affluent members of the population may have attended the same demonstrations as the February 20 movement supporters but for different reasons. They were less concerned with the ability to freely voice their opinions than they were with their ability to find work and feed their families. The constitutional reforms, particularly the food subsidies, may have been perceived by the lower classes as attempts by the monarchy to mitigate poverty, an issue of acute concern for the poor. They may view reforms as the king leading the way to help impoverished Moroccans, while corrupt officials (scapegoats) have absconded with the money intended to alleviate the plights of the poor. Issues of governmental accountability, civil rights, and freedom of speech, the aspects of the society with which bloggers and activists take issue, may not be concerns with which the lower classes can identify. Moreover, since the PJD has a less tarnished reputation, the majority of Moroccans may believe that the party can reform politics from within, obviating the necessity to completely demolish and rebuild the establishment.

7   Conclusion

In the case of Morocco, OCA through the Blogoma fails to reflect social and political divisions within the society, but it has contributed to a larger and freer civic sphere, because the bloggers may have aid the foundation for the development of offline activism to challenge the government. However, this space is still invisible to the average Moroccan who may be illiterate and is not likely to follow blogs in French. The average Moroccan may agree that corruption is apparent but accepts the scapegoats proffered by the makhzen and seems less interested in dethroning the monarch, who represents stability, consistency, and Moroccan Islamic identity. Joffe (2011) has argued that the Arab Spring did not topple the Moroccan monarchy due to Moroccan perceptions of him as a "liberalizing autocrat." Thus, average Moroccans are looking for ways to reform the system from within.

The flavor of reform that has been most popular addresses social and economic issues, and Islamic-oriented parties, such as the PJD, have offered social services in the absence of the government's ability or desire to do so (Cohen and Jaidi, 2006; Mekhennet, 2011). The poorer classes are most likely the greatest supporters of the PJD because, if the small upper/middle class holds similar views to those of the Blogoma, it wants nothing to do with the integration of religion into politics. Furthermore, the less affluent, typically coinciding with the more religiously conservative, are probably not too concerned with the PJD cracking down on liquor sales, segregating co-ed beaches, and encouraging more modest dress. The concerns of the Blogoma, on the other hand, might represent a powerful element of society that will continue to seek political liberalization from the monarchy and the advent of online social media to which these Moroccans have access represents a new realm for public discourse that the country has not previously known. It is possible that a future event could be framed to allow mobilization to occur within this fragmented society, as in the case of Tunisia (Lim, 2013). Time will tell if the unheeded calls for change will erupt into more violent reactions, causing the international community to exclaim that Morocco has molted its exceptionalism. The Arab Spring has not passed Morocco by—it is in hibernation for the winter.

References

Alami A (2012) Moroccan protests one year on. New York Times, February 15. Retrieved 17 Mar 2013, from http://www.nytimes.com/2012/02/16/world/africa/moroccan-protests-one-year-on. html?_r=0

Almiraat H (2011) Legislatives Marocaines Je Boycotte et Voici mes Raisons. Moroccan Mirror. Retrieved 9 Jan 2012, from Moroccan Mirror: http://almiraatblog.wordpress.com/2011/11/23/ legislatives-marocaines-je-boycotte-et-voici-mes-raisons/

Belhaj I (2009) Moroccan bloggers create first association. Maghrebia, April 30. Retrieved 19 Mar 2013, from http://www.magharebia.com/cocoon/awi/xhtml1/en_GB/features/awi/features/ 2009/04/30/feature-02

Ben Gharbia S (2007) Morocco: stop internet censorship! Global voices online, October 29. Retrieved 29 Apr 2008, from http://www.globalvoicesonline.org/2007/10/29/morocco-stop-internet-censorship/

Ben Gharbia S (2008) Morocco: Facebook's fake prince could face five years in prison. Global Voices Online, February 19. Retrieved 10 Mar 2009, from http://advocacy.globalvoicesonline. org/2008/02/19/morocco-facebooks-fake-prince-could-face-five-years-in-prison/

Benford R, Snow D (2000) Framing processes and social movements: an overview and assessment. Annu Rev Sociol 26:611–639

Bensalah M (2007) La BLOGOMA, histoire d'une reussite marocaine. Moi dans tous mes etat, November 21. Retrieved 29 Apr 2008, from http://moidanstousmesetats.blogspirit.com/ archive/2007/11/21/la-blogoma-histoire-d-une-reussite-marocaine.html

Cabalamuse (2011) PJD: mission impossible? A Moroccan about the world around him, November 29. Retrieved 12 Jan 2012, from http://cabalamuse.wordpress.com/2011/11/29/ pjd-mission-impossible/

Cherkaoui N (2008) Interview with Moroccan blogger Mohammed Erraji. Maghrebia, September 16. Retrieved 7 Mar 2009, from http://www.magharebia.com/cocoon/awi/xhtml1/en_GB/fea tures/awi/features/2008/09/16/feature-02

Cohen S, Jaidi J (2006) Morocco: globalization and its consequences. Routledge, New York

Couldry N (2003) Digital divide or discursive design? On the emerging ethics on information space. Ethics Inf Technol 5:89–97

Eatbees (2008) Moroccan injustice system. Eatbees Blog, February 26. Retrieved 3 Mar 2008, from http://www.eatbees.com/blog/2008/02/26/moroccan-injustice-system/

El Amrani I (2008) Free Fouad Mourtada. Arabist, February 15. Retrieved 3 Mar 2008, from http:// arabist.net/archives/2008/02/15/free-fouad-mourtada/#comment-392305

El Hilali L (2007) Les 100 qui font bouger le Maroc. Comme une bouteille jetee a la Mer, August 13. Retrieved 11 Mar 2008, from http://www.larbi.org/post/2007/08/13/420-les-100-qui-font-bouger-le-maroc

El Hilali L (2008a) En Defense de Fouad Mourtada. Comme une bouteille jetee a la Mer, February 11. Retrieved 11 Mar 2009, from http://www.larbi.org/post/2008/02/11/533-en-dfense-de-fouad-mourtada

El Hilali L (2008b) Fouad Mourtada condamné à trois ans de prison ferme. Comme une bouteille jetee a la Mer, February 22. Retrieved 11 Mar 2009, from http://www.larbi.org/post/2008/02/22/540-fouad-mourtada-a-t-condamn-trois-ans-de-prison#comments

El Hilali L (2011). Pourquoi je rejette la Constitution Mohammed VI. Comme une bouteille jetee a la Mer, June 18. Retrieved 28 Jan 2012, from http://www.larbi.org/post/2011/06/Pourquoi-je-rejette-la-Constitution-Mohammed-VI

Erraji M (2008). قصتي مع الكتابة والاعتقال : الفصيل الكاملة. Almassae, October 9. Retrieved 12 Mar 2009, from http://almassae.maktoobblog.com/1360512

Errazzouki S (2012) February 20th movement in retrospect: the treacherous path of reform. Alakhbar English, January 5. Retrieved 10 Jan 2012, from http://www.diigo.com/bookmark/http%3A%2F%2Fenglish.al-akhbar.com%2Fcontent%2Ffebruary-20th-movement-retrospect-treacherous-path-reform?tab=people&uname=hisham_almiraat

Etling B, Kelly J, Faris R, Palfrey J (2009) Mapping the Arabic blogosphere: politics, culture, and dissent. Berkman Center for Internet and Society at Harvard University, Cambridge, MA

Farrell H (2012) The consequences of the internet for politics. Annu Rev Pol Sci 15:35–52

Gamson W (1992) Talking politics. Cambridge University Press, New York

Gerhards J, Rucht D (1992) Mesomobilization: organizing and framing in two protest campaigns in West Germany. Am J Sociol 98:555–595

Graham M (2011) Time machines and virtual portals: the spatiality of the digital divide. Prog Dev Stud 11:211–227

Hogan M (2011) Social media v older Islamist revolutions. Aqoul, February 8. Retrieved 10 Jan 2012, from http://www.aqoul.com/archives/2011/02/social_media_v.php#comments

Ibahrine M (2004) Towards a national telecommunications strategy in Morocco. First Monday, 9. Retrieved 1 May 2008, from http://firstmonday.org/htbin/cgiwrap/bin/ojs/index.php/fm/article/view/1112/1032

Imad (2011a) (comments on Moorish Wanderer). Thank you your majesty, but no thanks. Moorish Wanderer, March 11. Retrieved 12 Jan 2012, from http://moorishwanderer.wordpress.com/2011/03/11/thank-you-your-majesty-but-no-thanks/#comments

Imad (2011b) (comments on Moorish Wanderer). Believe in something and dare everything. Moorish Wanderer, March 15. Retrieved 12 Jan 2012, from http://moorishwanderer.wordpress.com/2011/03/15/believe-in-something-and-dare-everything/#comments

International Telecommunications Union (2011) http://www.itu.org

Joffe G (2011) The Arab spring in North Africa: origins and prospects. J North Afr Stud 16(4):507–532

Klein G (2013) Two years on, Morocco's February 20 movement weakened. Middle East Online. Retrieved 17 Mar 2013, from http://www.middle-east-online.com/english/?id=57076

Lady Zee (2009) Don de la société civile marocaine: 25 Tonnes de médicaments arrivés à Gaza. Lady Zee, January 19. Retrieved 29 Jan 2012, from http://web.archive.org/web/20090124031140/http://ladyzee.wordpress.com/2009/01/18/don-de-la-societe-civile-marocaine-25-tonnes-de-medicaments-arrives-a-gaza/ (blog now defunct)

Lalami L (2008) Blogger Arrested in Morocco. Laila Lalami, September 9. Retrieved 1 Dec 2008, from http://lailalalami.com/2008/blogger-arrested-in-morocco/

Lalami L (2011a) Rocking the Casbah: Morocco's day of dignity. The Nation, February 20. Retrieved 10 Jan 2012, from http://www.thenation.com/print/blog/158751/rocking-casbah-moroccos-day-dignity

Lalami L (2011b) Morocco's moderate revolution. Foreign Policy, February 21. Retrieved 10 Jan 2012, from http://www.foreignpolicy.com/articles/2011/02/21/moroccos_moderate_revolution?page=0,0

Lim M (2013) Framing Bouazizi: 'White lies', hubrid network, and collective/connective action in the 2010-11 Tunisian uprising. Journalism 14(7):921–941

Lynch M (2009) Young brothers in cyberspace. Middle East Report Online, 245. Retrieved 19 Mar 2013, from http://www.marclynch.com/wp-content/uploads/2011/03/Middle-East-Report-245_-Young-Brothers-in-Cyberspace-by-Marc-Lynch.pdf

Maddy-Weitzman B (2001) Contested identities: Berbers, 'Berberism' and the state in North Africa. J North Afr Stud 6(3):23–47

Mamfakinch.com (n.d.) Retrieved 19 Jan 2012, from: http://www.mamfakinch.com/

Mekhennet S (2011) Moderate Islamist party winning Morocco election. New York Times, November 26. Retrieved 20 Mar 2013, from http://www.nytimes.com/2011/11/27/world/africa/moderate-islamist-party-winning-morocco-election.html?_r=0

Moroccans for Change (M4C) (2011a) #Feb20 Timeline – what next Morocco? Moroccans for Change, March 26. Retrieved 10 Jan 2012, from http://moroccansforchange.com/2011/03/26/feb20-timeline-what-next-morocco/

Moroccans for Change (M4C) (2011b) How Many #Feb20 movements are there Morocco? We've got a list. Moroccans for Change, March 31. Retrieved 10 Jan 2012, from http://moroccansforchange.com/2011/03/31/how-many-feb20-movements-are-there-morocco-weve-got-a-list/

Myrtus (2008) Hot topic. Myrtus Retrieved 3 Apr 2009, from http://www.myrtus.typepad.com/myrtus/2008/09/page/2/

Rahman Z (2012) Online youth politics activism in Morocco: Facebook and the birth of the February 20th movement. J New Media Stud MENA. Retrieved 18 March 2013, from www.slideshare.net/soussonline/online-activism-2012

Reporters Without Borders (RSF) (2008) Court releases blogger who criticized king. Reporters without Borders, September 11. Retrieved 23 Mar 2009, from http://www.rsf.org/article.php3?id_article=28449

Roumana (2011) Cult of personality and other thoughts on Moroccan Monarchy. Kobida, February 6. Retrieved 12 Jan 2012, from http://kbida.wordpress.com/2011/02/06/cult-of-personality-and-other-thoughts-on-moroccan-monarchy/

Selwyn N (2004) Rethinking political and popular understandings of the digital divide. New Media Soc 6(3):341–362

Soutienerraji (Free Erraji) (2008) Mohamed Erraji : La mobilisation continue/ERRAJI STILL FACES 2yrs IN JAIL. YOU CAN HELP. soutienerraji. Retrieved 20 Mar 2009, from http://soutienerraji.blogspot.com/

Tourabi MA (2011) Une classe tres moyenne. Moroccans for Change, March 14. Retrieved 12 Jan 2012, from http://moroccansforchange.com/2011/03/14/excellent-piece-feb20-a-very-middle-class-translated-from-une-classe-tres-moyenne-by-m-abdellah-tourabi/

Warschauer M (2004) Technology and social inclusion: rethinking the digital divide. MIT Press, Cambridge, MA

York J (2008) Morocco: bloggers react to Fouad Mourtada's release from prison. Global Voices, March 19 Retrieved 26 Apr 2008, from http://www.globalvoicesonline.org/2008/03/19/morocco-bloggers-react-to-fouad-mourtadas-release-from-prison/

Online and Offline Advocacy for American Hijabis: Organizational and Organic Tactical Configurations

Rebecca S. Robinson

Abstract This paper outlines the divergent tactics of two groups of American Muslim collective actors, which are conceptualized in terms of organizational and organic, in addressing the issue of hijab and discrimination in the workplace. The organizational form, represented by the Council for American-Islamic Relations (CAIR), uses highly publicized awareness campaigns to promote civil rights, while the organic form, represented by the hijabi fashion community, tends to opt out of drawing attention to the discrimination that is commonplace in their dealings with other Americans. The term organic is employed to demarcate the differences between structured grassroots and nonhierarchal collective action carried out through the Internet. A qualitative content analysis of seven hijabi fashion blogs is conducted to cull thematic continuities among these bloggers and their commentators in regard to a few well-publicized campaigns against hijab discrimination launched by CAIR and other discussions of discrimination in workplace. The paper argues that the tactics of the organic form of collective action develop discursively through discussions on these websites about how hijabis can navigate various situations, grounded in a shared sense of identity and context. These tactics are customized for experiences of hijabis "on the ground" as opposed to the organizational tactics, which model the successes of other civil rights organizations. The paper concludes by suggesting that the organizational tactics might be more effective, in the context of the continued expansion of Islamophobia in the American culture, if American foreign policy and mainstream media did not continue to promote the connection between Islam and extremism.

R.S. Robinson (✉)
Arizona State University, Tempe, AZ, USA
e-mail: Rebecca.Robinson@asu.edu

N. Agarwal et al. (eds.), *Online Collective Action*, Lecture Notes in Social Networks, DOI 10.1007/978-3-7091-1340-0_12, © Springer-Verlag Wien 2014

1   Introduction

CAIR, the Council for American-Islamic Relations, is a nonprofit organization
dedicated to promoting the rights of all Americans but focuses particularly on
furthering the rights of American Muslims. Similar to other civil rights organizations, CAIR encourages Muslim solidarity through engendering a collective identity and advocating for the rights of Muslim Americans. The organization publicizes
cases of infringement on the rights of Muslim Americans, such as workplace
discrimination, which will be the focus of this paper. Unlike the social norm of
laïcité in France, in the United States, Americans ostensibly have the right to
express their religious beliefs in public spaces, including public schools and places
of employment. Nevertheless, Muslim women in hijab, the Muslim headscarf, often
face discrimination due to their conspicuous religious apparel. One might expect that
the discrimination to which hijabis, women in hijab, are subjected would enhance
their solidarity, but counter evidence was observed.

While conducting the content analysis for a paper on American sexuality and
hijab, several comments were found on a popular American hijabi fashion blog,
HijabTrendz, about the case of Imane Boudlal being suspended from her job as a
hostess at a Disney restaurant for refusing to remove her headscarf. The comments
of the blog's author, Miriam Sobh, and other interlocutors on the blog were largely
unsupportive of Boudlal's plights (Sobh 2010).

Boudlal's suspension was taken up by CAIR, but the fact that many commentators on the hijabi fashion blog expressed more solidarity with Disney than with
a fellow Muslim sister suggests that perhaps there was a schism between the
tactics that the civil rights organization employs to promote the rights of Muslim
Americans, hijabis in particular in this case, and how the population it was
attempting to represent believes that it can best protect the rights of hijabis.
In other words, some of the hijabis did not appear to favor the protection of
Boudlal's workplace right or the amount of notoriety the case was receiving
due to CAIR's campaign.

This paper will take up the issue of collective action tactics and how organizational actors' strategies might differ or contradict those of organic, online
collectives. The chapter will begin with a review of literature connected to collective action tactics. Secondly, it will describe some of commonalities and differences
between CAIR and the hijabi fashion blog community, mostly in relation to how
they frame their movements and their target populations. Thirdly, it will attempt to
explain the discontinuities related to their tactics in promoting the rights of hijabis
in the workplace. Finally, the chapter will offer some concluding remarks about
why the tactics of the organic collective actors make intuitive sense in the context of
Muslims living in the United States, despite the obvious importance and necessity
of CAIR's mission and efforts.

2   Framing Organic and Organizational Collective Action

It is necessary to begin by conceptualizing the terms organizational and organic. Both groups of collective actors broached in this chapter are grassroots movements because they arose to address the stigmas attached to a minority population in the American culture. CAIR addresses civil rights abuses and discrimination, while the hijabi population emphasizes their exclusion from the popular American culture in terms of the limitations of modest dress options and the prejudice they endure because of their hijabs. Grassroots initiatives gain momentum spontaneously through community mobilization, but there are evident distinctions between hierarchically structured organizational configurations and self-organized, networked forms of mobilization.

The conceptualization of organic used here draws upon the theory of "organic solidarity" outlined by Durkheim. Durkheim employed the term organic solidarity in regard to less hierarchical configurations involved in the division of labor as societies evolved into more complex forms (Durkheim 1984), but the terminology can be applied to other situations in which independently functioning interlocutors spontaneously engage in interaction without organizational constraints. Although there may be more or less influential members in organic collectivities, there is no predetermined leadership, mission, or top-down directives in regard to the tactics employed by the actors. The organic configuration persists simply because of the mutual benefit it offers to the collective actors involved.

However, one may question if it is possible for a loosely configured, organic collective to actually utilize collective tactics, and why it is important to consider these tactics in the first place as opposed to other aspects of collective action, such as frames and collective identity. Since the main distinctions between the two forms of collective action discussed here are their structural configurations, organic and organizational, and the tactics that they use to address the problematics of hijabs in the workplace, the exploration of collective action tactics is of central importance to this chapter.

Wilson (1973) claims that movements are often remembered more for their tactics than their missions. Jasper (1997) stresses the importance of tactics through his discussion of the major distinctions between forms of collective action: activist, organizational, and tactical. The tactical repertoires of collectivities not only inform the ways in which collective actors mobilize, recruit, and protest but they oftentimes relate to the identities of the collective actors and self-representations of the movements. Collective actors may share similar identities, advance comparable frames, and address the same audiences but be viewed as highly divergent simply in relation to their preferences of tactics.

The relative radicalism or moderateness of the collective action impacts the organizations' or groups' acceptability in the mainstream and, therefore, the extent to which their diagnostic and prognostic frames are integrated into the beliefs of the larger society. For example, McAdams (1982) points out the mainstream acceptance of nonviolent activism during the Civil Rights Movement lessen the effectiveness of more militant strategies proposed by other groups, like the Black Panthers. Similarly, environmental movements, such as the Sierra Club and Earth First!, both aim at promoting environmental protection, but their strategies contribute to how they are perceived by prospective recruits and the mainstream. The campaigns conducted by Earth First! are often considered too radical to be accepted by the mainstream society. Hence, the legitimacy of collective actors is strongly connected to how they execute their collective action. Both groups examined here ascribe to moderateness in their political and religious philosophies; however, this may not ultimately sway their legitimacy in the eyes of mainstream Americans because affiliation with Islam is often perceived as synonymous with extremism.

The perceived extremism associated with Islam and its widespread rejection by the mainstream make collective action through the Internet more attractive to some Muslims in terms of self-preservation. Two issues are particularly relevant to hijabis. Firstly, Chong (1991) claims, in reference to civil rights movements, that part of the success of collective action has to do with the extent to which the tactics of collective actors lend to the notoriety of the movement. However, if the intent of the collective action is not to gain notoriety but to form a support network, the component of notoriety is unnecessary—hijabis tend to get enough unwanted attention without affiliating with a movement. It is also possible that when collectives form online, one of the overarching aims for doing so is to allow them to remain less noticeable. Secondly, McCarthy (1987) claims that prospective collective actors that lack infrastructure and opportunities for everyday interaction can develop collective identities, which facilitate network ties. These ties can blossom into communities through the Internet, forming the "counterpublics" advanced by Fraser (1992). Though it is the fastest growing religion in the United States, considering the scarce number of Muslims in general, this type of hijabi network would be much less likely to develop without the Internet. Hijabis interested in modest fashion would not likely be able to find mutual support in their local communities, even through the local masjid, so they go online to find it.

The hijabi fashion community can best be characterized as a support network, similar to those theorized by Smith and Stevens (1999), in which hijabis engage with each other to mitigate the stress associated with the rampant Islamophobia in the American culture. However, returning to the question as to how the hijabi collective actors formulate their tactics, the network serves the dual purposes of lending support and the space to discuss how to navigate more pragmatic issues. Through sharing their realities with other hijabis, they are able to reaffirm their experiences with discrimination despite the supposed tolerance of America, so they gain a better sense of the context in which they live. After they develop this shared sense of context, they are able to discursively explore tactics for dealing with various situations. For example, a question posed on one of the websites was whether or not there were work options that would be unsuitable for hijabis— jokingly, some women suggested Hooter's might not be a practical employment option (Sobh 2009). While a comprehensive consensus on issues related to the hijab in the workplace did not develop within these discussions, their strategies for dealing with discrimination stood in stark contrast to CAIR's tactics.

The theoretical position of this paper is that, though the identities, frames, and target populations of the organizational (CAIR) and organic (hijabi fashion community) movements converge on multiple levels, the tactics that they employ are markedly distinct. Perhaps the tactical differences can be allotted to the experiences of those "on the ground" in contrast to an organizational form that has modeled similar longstanding organizational forms. The organizational form promotes civil rights through the widespread publicizing of Islamophobia and discrimination, while organic collective actors recognize these rights but tend to opt out of confrontational tactics to protect themselves. Their tactics can best be characterized as flight rather than fight. Their everyday experiences lend to their conviction that fighting may encourage backlash against Muslim, and they are the most visible targets because of their hijabs.

3 The Tactic Is the Technology

The hijabi movement demonstrates that collective action through the Internet can move beyond the instrumental and symbolic framings of it presented by Nip (2003). In this case, Internet use provides a space where hijabis can interact freely, a space free of the disparagement they face in the offline universe. While the space serves symbolic functions, in that the collective actors strive to empower members and develop an overarching sense of collective identity as well as parameters of inclusion and exclusion, it also performs a tactical purpose. Similar to *hacktivism*, the technology itself is central to their tactical repertoire; their tactical withdrawal from the larger society to protect themselves is provided through the technology. In contrast to theorists who opined in the early stages of the Internet's development that it would bring about social isolation in a negative sense (Katz and Rice 2002), this form of isolation is fundamentally sociable and collective.

4 Methods

The author conducted a grounded, qualitative content analysis of seven hijabi fashion blogs (HijabiTopia, HijabTrendz, We Love Hijab, Fashioning Faith, Modesty Theory, Luff is All You Need, and Haute Muslimah) to explore thematic continuities among the bloggers and commentators on these blogs. The blogs made up a network based on their content and the bloggers' commentary on each others' blogs. Many of the bloggers also participate in a bi-annual weeklong initiative started by "Em" of Modesty Theory—Hijab Fashion Week (Em n.d.).

Posts between February 2010 and June 2011[1] were included in the analysis. The principal goal of this content analysis was to discover how the hijabis reacted to the Boudlal and Khan cases and other postings and commentary related to discrimination they have faced in the workplace and how they believed that the hijab might affect their employability. The "About" pages of the blogs were also reviewed to gain a greater understanding of the bloggers' motivations for blogging toward elucidating the frames of the movement.

CAIR's website was also examined to understand how it frames the issue of Muslim American civil rights and the tactics it employs to promote greater understanding of Islam, the rights of Muslim workers, and the normalization of the American Muslim identity in mainstream discourse.

5   Target Populations

The frames and the target populations of the hijabi fashion movement and CAIR converge on many levels. In regard to target populations, many of CAIR's campaigns call upon Muslims and "people of conscience" to speak out on various issues through letter writing and spreading awareness; thus, it hopes to influence the entire American population. American Muslims, however, stand to benefit the most from the normalization of Muslim identities and Islam through these campaigns; thus, the American mainstream is the focus of CAIR's campaign and American Muslims are the population that stands to benefit the most.

Similarly, some of the hijabi fashion bloggers want to build a network that supports the decision of hijabis and other American women to dress modestly. Some also mention wanting to impact the fashion industry and/or countering the commodification of women's bodies on a larger scale. Therefore, their target readers are Americans who are interested in modest fashion or diminishing the misogyny associated with the mainstream fashion industry. Similarly, however, other hijabis would tend to benefit the most from the normalization of the hijab and modesty in the American mainstream.

6   Frames of the Hijabis Fashion Community

The hijabi fashion bloggers frame the movement in various ways but they often refer to bringing modesty into fashion, forging a community, and creating positive role models. Many of the bloggers highlighted here have linked the creation of their blogs to the paucity of material options for modest fashion in the American society

---

[1] Khan was terminated in February 2010 and filed a lawsuit in June 2011. Boudlal was suspended in August 2010.

and state that their motivation for blogging is to promote Islamic-appropriate dress, modest fashion options, and pride in the American hijabi identity. They blog to share ideas about designing couture that is both fashionable and modest, often using runway ensembles and altering them to comply with their standards of modesty.

Creating a space where hijabis can interact seems to be one of the most important motivations for hijabi fashion blogging. One commentator mentioned that part of the reason that she decided to wear hijab was because of the "We Love Hijab" website: she writes, "knowing there is a group of woman i can turn to for support was very positive for me" (Hamidullah 2010). Another states, "Although I am now growing more comfortable in my hijab, I do feel that many people tend to stare at me...I find that 'forums' like this give a sense of support for girls like me who do at first struggle with criticism..." (ibid). In an article that features Sobh of HijabTrendz, she is quoted in saying, "With the Internet, there's so much out there. There's so much support" (Torres 2010). Diana, writing for *Muslimah Media Watch*, states, "I...feel as though *I am not autonomous, because other people have already spoken for me*...Muslim women have the wherewithal and intelligence and...the space to speak for themselves" (Diana 2010). The last comment highlights the important concern of Muslim women that they are often denied a vehicle through the mainstream media to speak for themselves and the Internet has provided them with this opportunity.

This space for hijabis to interact is very important because, while wearing the hijab is often considered a personal struggle (jihad) to bring one closer to Allah and/or to acquiesce with the mandates of the Qur'an, the knowledge that other women endure similar experiences is a comfort. Jokima Hamidullah of "We Love Hijab" claims, "wearing hijab can be a very stressful experience outside of the community" (Hamidullah 2010). "just like issues of skin colour prejudice or straight hair vs. kinky hair -being different is still a major issue. And women always get it worse on these things," says an interlocutor on this blog (ibid). Another commentator says, "You can't work if you [are] wearing hijeb, you can't go to school...You can't have a normal life... it [is] consid[e]red like a soumission [slave[ry]] for women..." (ibid).

To counter this connection between "slavery" and the hijab, some bloggers cite the need to develop positive hijabi role models. Stephanie Luff, who blogs at "Luff is All You Need," claims she wants to act as one of these. She claims that modestly dressed role models are severely lacking in the American culture and "girls starts to watch these western role models & believe that the immoral lifestyle is a suitable alternative" (Luff 2010). On the "Modesty Theory" website, when Em writes about her decision to no longer model her fashion ensembles, one commentator suggests that Em's decision is related to her recognition that, as a blogger, she is an example to others (Em 2010). Some of the websites, like HijabiTopia and Haute Muslimah, showcase various "Super Hijabis" (hijabis that deserve recognition for their accomplishments) and hijabis that have been successful in the modest fashion industry as role models. A few women state that wearing the hijab is "paving a way" for other Muslim women to do the same (Hamidullah 2010).

As demonstrated above, the frames of the hijabi fashion community vary substantially, particularly in comparison to a unitary, hierarchical organization. The discussion of how CAIR frames its mission will be brief because this information is readily available on the organization's website, but it functions similarly to other civil rights organizations. CAIR was established in 1994 and claims to be the largest Islamic grassroots organization in the United States (CAIR 2012) with 29 chapters across the country. Its mission is to "enhance understanding of Islam, encourage dialogue, protect civil liberties, empower American Muslims, and build coalitions that promote justice and mutual understanding" (ibid). It advances its mission through various humanitarian, advocacy, and educational campaigns.

7 Confluence of Frames

While CAIR perhaps takes a conciliatory approach in attempting to normalize perspectives of Islam in America, some of the hijabis demonstrate increasing frustration with having to tolerate the insensitive remarks, absurd questions, and blatant derision that are commonplace in their dealings with other Americans. When a drama instructor questioned whether Sobh had difficulty hearing because her scarf was too tight, many hijabis responded with similar experiences (Sobh 2011a). A few questioned how long hijabis would be forced to endure the discrimination they often face (ibid). Another suggested that it was acceptable to be sensitive about inappropriate comments like the one her instructor made (ibid). Several hijabis claim to have been asked questions about wearing the hijab in the shower or whether or not they get hot (Parvez 2011; Kavakci 2010). One commentator claims that a man called her a "shariah lawyer" when she complained about the fumes from his generator (Sobh 2011a). One hijabi stated that she did not want hijabis to put up with derogatory comments in an attempt to remain "above it all" (ibid). Other hijabis discussed their experiences in the workplace, which will be broached presently.

In contrast to those hijabis that have become inured to the prospects of the normalization of their identities, a few hijabis, similar to CAIR's advocacy, believe that they can influence the perceptions of non-Muslim Americans. One commentator writes, through hijabis "people can better understand Islam!" (Hamidullah 2010). Another stated that hijabis should serve as ambassadors for Islam (ibid). Clearly, there is some hope that perceptions of Muslim women can be destigmtized in the mainstream.

Other hijabis were less hopeful. For some, the problem was not so much the belittling comments as the assumptions and prejudices. One woman claimed that non-Muslims assume that hijabis do not speak English or that they are uneducated (Hamidullah 2010). Sobh outlined four common assumptions that non-Muslims have about hijabi: "they are very religious, they are very serious, they do not have interests outside of having children, and they are not real Americans" (Sobh 2011b). Women in hijab are aware of the stigma attached to their identities, but their sentiments as to whether or not they are able to mitigate the prejudices of mainstream culture are somewhat ambiguous.

These comments about the exclusion of hijabis from those who can be considered legitimate Americans are relevant to CAIR as well. CAIR obviously believes that its campaigns have been or will be effective in influencing mainstream perceptions of Muslims, because its campaigns have remained consistent over the nearly two decades that it has been in operation. However, a HijabTrendz posting indicates the negative perspectives about Muslims persist and have actually worsened since 2001—anti-Muslim sentiment reached an all-time high in 2010 (Sobh 2011c).

Despite the pervasive Islamophobia in the American society, hijabis and CAIR express a great deal of pride in their Americanism. Sobh posts in jest a list of the "top 5 ways to announce to the world that you're Muslim": one is "Carrying an American flag at a rally but shouting in Arabic" (Sobh 2011d). Luff showcases her fashion ensemble for July 4th—dressed from head to toe in red, white, and blue. An interlocutor on "We Love Hijab" sums up her perception of religious freedom in the American context: "[in the United States] you have the freedom of religion, I love that and am so proud of that because I have the right to freely chose to cover my head" (Hamidullah 2010).

Likewise, to demonstrate its Americanism, on the "25 Facts About CAIR" webpage, the organization discusses some of its advocacy campaigns that are or have become part of mainstream American ideology: CAIR's opposition to the invasion of Iraq in 2003; CAIR's advocacy against the use of torture in interrogation; and CAIR's early rejection of the Patriot Act (CAIR n.d.a, 25 Facts). This section of the website is perhaps related to the accusations of anti-Americanism that are hurled at the organization, which will be discussed promptly.

Both CAIR and the hijabi community also stress their moderate adherence to Islam, which is arguably the norm of the Muslims both in the United States and abroad. Islam is a religion that preaches moderation. Some of the hijabis draw attention to the mainstream connection between the hijab and extremism. One commentator mentions that she was considered "extreme," excluding her from the ranks of moderate Islam, by her own Muslim family due to her decision to wear the hijab (Hamidullah 2010). Likewise, Kavakci (2010) emphasizes, "I'm not an extremist Muslim," refuting this common connotation of the hijab in mainstream America.

Similarly, CAIR attempts to refute the Islamophobic rhetoric that characterizes it as an entity that spreads anti-Americanism. One of the pages on its website attempts to dispel the "myths" associated with the organization: CAIR has backing or is connected to Wahhabism and/or other "fundamentalist" groups like Hamas, Hezbollah, and/or the Muslim Brotherhood; CAIR promotes anti-American extremism; and former associates of CAIR have been deported for terrorism-related charges (CAIR n.d.b, Disinformation). In refutation of these allegations, CAIR writes that it condemned the 9/11 attacks within the first few hours of the incident and, later, issued a public service announcement, entitled "Not in the Name of Islam," which attempts to demystify the connection between Islam and terrorism, and a fatwa to declare the murder of civilians as haram (forbidden) (ibid). CAIR has also been accused of subverting the U.S. Constitution, although it claims that most of its advocacy is based on upholding it (ibid). It has earned its unsavory reputation, not through its own workings but through its connection to Islam and the connection in mainstream American consciousness between Islam and extremism.

Neither CAIR nor some hijabis question the incongruence between their belief in the tolerance of American society and the substantial discrimination they face as "others" in the culture. One commentator demarcates the difference between discriminatory legislation and personal prejudices against Muslims, citing the former as more insidious (Hamidullah 2010). While it is difficult to gauge the extent to which Muslims agree with this public/private dichotomy of prejudice, the fervent Americanism of some hijabis and CAIR, in the face of the pervasive Islamophobia, speaks to the somewhat widespread acceptance that discrimination is due to the private intolerance of ignorant people rather than a public, systemic issue. Mohanty (2003) might say that these Muslims have bought into the fallacy of multiculturalism in the United States, which tends to pass off racist attitudes as personal defects rather than casting them as systemic ones, grounded in the country's foundations in white, Christian, heterosexual, able-bodied, male privilege.

8   Hijab in the Workplace

The hijab was understandably a concern for some of the interlocutors in the hijabi community who were in the workforce. One commentator claims that the first time she wore the hijab to work after converting her boss took her into a different room, "slammed his fist on the table and said, 'If you don't take off that da_n scarf, you won't be working here'...I'd always done a good job...After that, I couldn't do anything right..." (Sobh 2011a). Another hijabi had a supervisor that told other employees that she was oppressing herself and allowing men to dictate her life (ibid). Another interlocutor mentions, having spoken to a "number of working and professional Muslimahs," "...the experience of stares, hostility and aggression b/c of wearing the hijab is felt all around" (Hamidullah 2010). Another woman agrees with this comment, stating that workplaces are hostile to "outward expressions of religion" (ibid).

This is not to say that none of the hijabis had positive experiences in their places of employment, but the reservation of non-hijabis to put on the hijab and look for a job was tangible, even for those who believe that it is religiously mandated. When hijabis invoked disapprovingly the excuses that non-hijabis make for not wearing the hijab, employment options were typically at the top of the list, even though many hijabis agree that the hijab definitely impacts employment options (Sobh 2010; Hamidullah 2010).

CAIR does not believe that the hijab should affect the employment options of Muslims, so it has created a pamphlet, entitled "An Employer's Guide to Islamic

Religious Practices," to disseminate information about the rights of Muslims in the workplace (CAIR 2005). One section that focuses specifically on women suggests that some Muslim women may wear headscarves or face veils and, although employers may ask people who wear religious apparel to use certain colors or fabrics, they should consider modifying their dress codes to accommodate religious apparel. The pamphlet specifically cites altering "no hat" policies.

Some hijabis understand their rights as outlined by CAIR, that is, employers must modify their regulations to accommodate religious beliefs. One says, "they can not do anything to you. If they fire u because of [wearing hijab], they should be ready for a law suit" (Hamidullah 2010). Despite this commentator's apparent understanding of the rights guaranteed to her under the 1964 Civil Right Act in regard to religious accommodation, there was less evidence that these rights were widely acknowledged within the hijabi community.

The case of Boudlal's suspension from her job at Disneyland was illustrative of the misunderstandings associated with religious accommodation in the community. One hijabi writes, "If a woman came in and said "I only want to wear a sleeveless shirt" and she refused to wear the proper uniform she would be fired for not following the rules" (Sobh 2010). Another suggests that corporations are "very sensitive" about their employees upholding a particular "image" (ibid), although the ADL website stipulates that preventing employees from wearing religious garments is limited to "safety issues" and employers cannot ask employees to remove these garments due to the potential discomfort of their clientele (ADL n.d.). A third commentator says, "I'd rather see a company offer a compromise than fire her for her beliefs" (Sobh 2010). A fourth woman states, "I also wear hijab and i had to leave my job because of it...when you think how it's difficult to find a job...employers have so much choice...they are not forced to propose a compromise for religions reasons" (ibid). While the first two comments demonstrate a lack of understanding between failure to comply with a dress code and failure to comply with a dress code for religious reasons, the latter comments tend to view religious accommodation in an optional sense rather than as a right that is mandated by federal law.

Blatant discrimination and termination after being hired appeared to be as much of a concern for some interlocutors as not being able to get a job in the first place. Sobh, who started the thread about Boudlal's case, states, "it's pretty cool to have a company that is willing to give you an alternative...most people...wou[l]d basically never offer you a job" (ibid). Another commenter claims, "Many places don't even hire you because of wearing hijab at least not me" (ibid). A third claims, "many people nowadays can't get a job...because of their religious beliefs" (ibid). Even though some hijabis acknowledge the fact that it is exorbitantly difficult to secure employment in the hijab, they were more sympathetic of Disney's compromise, which it is legally mandated to offer, than of Boudlal's desire to wear the hijab on the job. Despite a few comments indicating that Disney had only offered an alternative *after* Boudlal had filed suit, though she had waited to hear back from the company about her request for religious accommodation for months, few commentators believed that Disney's actions were discriminatory. The women that favored

Disney's alternative ignored the comments about it being offered after Boudlal had requested assistance from CAIR in filing the EEOC complaint.

Boudlal rejected Disney's alternative because she said it was silly and mocked her religious beliefs. A few also agreed that the alternative was "silly" (ibid) and "a bit overboard" (ibid) and that Boudlal would look out of place anyways, either in hijab or in Disney's proposed alternative, because she would be the only employee wearing "headgear." However, the woman who thought Disney's alternative looked "silly" claims that she would accept a Disney job that was out of the view of the public or she would quit (ibid) rather than fight for her right to exercise her religious beliefs. Several hijabis believe that Boudlal's rejection of Disney's compromise reflects poorly on other hijabis and might preclude employers from offering alternatives in the future, again, demonstrating that they believe it is an option rather than the law.

This suggestion to work outside of the public eye evokes the question as to whether or not there are more or less appropriate places for hijabis to seek employment. In regard to the Boudlal case, "as a muslima, i couldn't even work for a firm like disney (due to, among others, their implication with israel)," writes one woman. The hijabi community was much more sympathetic to Khan's experience, because, unlike Boudlal, she had interviewed and been hired in her hijab. Even so, Kavakci, blogging at HijabiTopia, claims that Muslim women should be careful about where they seek employment: "... Muslim girls please choose where you work wisely. A store that greets its customers with half naked model looking guys, is not a place for a muslimah to work! There are so many places for a muslim girl to work at." Her statement denies the perspective of several other hijabis that it is very difficult to get hired in a hijab. Further, contrary to Kavakci's assertion, Zahra Billoo, the CAIR Outreach Director in the San Francisco Bay Area, the branch that took up Khan's case, says, "It's not ok that Muslim women have to think twice about where they can work, that their religious freedom is being impinged upon by employers who have blatantly decided not to follow the law" (AP 2010).

9 Tactical Repertoires

Both CAIR and the hijabi network attempt to have their worldviews accepted by the larger American population. The legitimacy of CAIR's tactics is couched in modeling other civil rights organizations, which rely on increasing awareness and highlighting cases of abuse through the extensive publicity of their campaigns; however, CAIR has become the target of attacks due to its high degree of notoriety in the mainstream. To counter these attacks, CAIR emphasizes its positions on various issues that have been accepted by the majority of Americans and/or reaffirmed by city/state legislation and the courts. The organization appears to believe that its strategies are effective and that its campaigns will promote change despite the continuous growth of Islamophobia in the society.

Similar to CAIR's ideology, hijabis tend to be proud, moderate Americans but, in contrast to CAIR's very public campaigns, evidence suggests that the tactics hijabis employ are related to maintaining a low profile. Some hijabis seem less convinced that the American society is moving toward greater tolerance of Muslims. One states that it is "en vogue" (Sobh 2010) to be intolerant toward Muslims, and another claims that this intolerance will endure "as long as the cold war" (ibid). Another hijabi points out that judges are more inclined to favor the right of businesses to enforce dress codes than to rule in favor of individual acts of discrimination against hijabis (Sobh 2010). While quite a bit of continuity exists among the frames and target populations of the organizational and organic collectives, the tactics they employ in regard to hijab in the workplace are quite polarized.

CAIR's campaign to improve the image of Muslim Americans involves drawing discriminatory action into the public sphere. However, CAIR itself recognizes that the high publicity of its campaigns has made it the target of anti-Islamic sentiment: "Because of CAIR's high profile...a small but vocal group of anti-Muslim bigots has made CAIR the focus of their misinformation campaigns. Internet hate sites then recycle these attacks..." (CAIR n.d.b, Disinformation).

Perhaps hijabis believe they have been on the receiving end of enough attacks due to the conspicuousness of their association with Islam so, rather than facing more attacks, they are inclined to withdraw from the public sphere. This tactic is not so much imposed upon the community in a top-down fashion, but it is a strategy that has developed discursively through the shared experiences of hijabis in an effort to protect themselves. Although the analysis of these blogs reveals the inclinations of a relatively small community, a related trend has been noted elsewhere in a larger sample of over 200 hijabis. According to Ghumman and Jackson (2010), hijabis tend to look for jobs which are outside of the public eye because they are aware of the stigma attached to their religious apparel. Therefore, despite the assertion of CAIR's spokesperson that hijabis should be able to work wherever they want because their religious beliefs are protected by law, the perception of hijabis is that it may be difficult to secure employment in hijab if they are not careful about where they apply to work or how they frame their identities (King and Ahmed 2010). Furthermore, if they face discrimination in the workplace because of their hijabs, they feel like they are better off looking for another job rather than standing up for their constitutional rights.

10   Conclusion

Both the organizational and the organic forms of collective action employ tactics that speak to their contextualized understandings of opportunities for American Muslims. CAIR wants to eradicate discrimination through creating widespread awareness about the pervasiveness of Islamophobia in the American society. Hijabis are more inclined toward self-preservation through withdrawing from the public eye and relying on the Internet to socialize in a safe place if they live in communities with few Muslims. While their tactics are diametrically opposed, neither one can be said to be superior—they are simply products of the various realities of Muslims living in America. However, the organizational collective actor may be less attuned to the "on the ground" experiences of the most visible targets of Islamophobia, the hijabis.

In the current political context of American foreign policy in the MENA region and the anti-Muslim current integrated into mainstream culture, it is difficult for CAIR to maintain legitimacy as an organization that is founded on moderateness and constitutional principles, when it is associated with a religion that is synonymous with extremism in the American imaginary. When a sizeable portion of the U.S. military budget is devoted to "maintaining political stability" in MENA nations and "opposing factions of militant Islamists," it is very easy for non-Muslim Americans, with the help of the mainstream media that too often presents Islam as a religion of extremists, to equate typical, moderate Muslims with fanatics. My point is that CAIR's mission may be untenable until there is a radical reimagining of Islam in the U.S. culture.

The tactics of the hijabis are customized for this context. They are disabused of the tolerance of the American people and few are persuaded that there is hope for the normalization of their identities in the foreseeable future. They tend to reject the high notoriety of the CAIR campaigns because these campaigns may actually encourage backlash against Muslims. They retire to an online space where they are safe and can feel empowered among their peers, awaiting a day when non-Muslim Americans gain a more balanced perspective of Islamic beliefs, and they will no longer have to justify their right to exercise their beliefs as they see fit.

Works Cited

ADL (n.d.) Religious accommodation in the workplace: your rights and obligations. Retrieved 23 Dec 2011, from Anti-Defamation League. http://www.adl.org/religious_freedom/resource_kit/religion_workplace.asp

Associate Press (2010) Muslim: 'Abercrombie Fired Me Over Headscarf' (http://www.youtube.com/watch?v=IOXJ2aoCn3w&feature=player_embedded). Retrieved 23 Jul 2012, from HijabTrendz. http://www.hijabtrendz.com/2010/02/25/hijabi-fired-from-abercrombie/

CAIR (2005) An employer's guide to Islamic religious practices. Retrieved 26 Dec 2011, from Council on American-Islamic Relations. http://sun.cair.com/Portals/0/pdf/employment_guide.pdf

CAIR (2012) Our vision, mission and core principles. Retrieved 26 Dec 2011, from Council on American-Islamic Relations. http://www.cair.com/AboutUs/VisionMissionCorePrinciples.aspx

CAIR (n.d.a) 25 Facts about CAIR. Retrieved 26 Dec 2011, from Council on American-Islamic Relations. http://www.cair.com/AboutUs/25FactsAboutCAIR.aspx

CAIR (n.d.b) Top internet disinformation about CAIR. Retrieved 26 Dec 2011, from Council on American-Islamic Relations. http://www.cair.com/Portals/0/pdf/Dispelling_Rumors_about_CAIR.pdf

Chong D (1991) Collective action and the civil rights movement. University of Chicago Press, Chicago, IL

Diana (2010) Coverage of "Fashionable" Muslim women cramps our style. Retrieved 28 Dec, from Muslimah Media Watch. http://www.patheos.com/blogs/mmw/2010/07/coverage-of-fashionable-muslim-women-cramps-our-style/ (linked through Fashioning Faith: http://www.fashfaith.com/2010/07/muslimah-fashion-in-media-what-do-you.html), 19 July

Durkheim E (1984) The division of labor in society. Free Press, New York, NY

Em (2010) Blog Redo Rant >_>. Retrieved 18 Jul, from Modesty Theory. http://modestytheory.blogspot.com/2010/09/blog-redo-rant.html, 27 September

Em (n.d.) About. Retrieved 18 Dec, from http://modestytheory.blogspot.com/p/about.html

Fraser N (1992) Rethinking the public sphere: a contribution of the critique of actually existing democracy. In: Calhoun C (ed) Habermas and the public sphere. MIT Press, Cambridge, MA, pp 109–142

Ghumman S, Jackson L (2010) The downside of religious attire: The Muslim Headscarf and expectations of obtaining employment. J Organ Behav 31:4–23

Hamidullah J (2010) Hot topic: who's got a problem with hijab? Retrieved 19 Jul 2011, from We Love Hijab. http://welovehijab.com/2010/02/26/hot-topic-whos-got-a-problem-with-hijab/, 26 February (Also see the comments of the following: mira, Maria, Samira, Shaday, Rachida, Paulette, mermz, Amy, Farisha, Carib Muslimah, Eleni, Yusra, tanya farook, Jana, Zehra)

Jasper J (1997) The art of moral protest. University of Chicago Press, Chicago, IL

Katz A, Rice R (2002) Social consequences of internet use: access, involvement, and interaction. MIT Press, Cambridge, MA

Kavakci E (2010) To wear or not to wear: it's not a question! Retrieved 17 Jul, from HijabiTopia. http://hijabitopia.blogspot.com/2010/08/to-wear-or-not-to-wear-its-not-question.html, 17 August

King E, Ahmed A (2010) An experiment field study of interpersonal discrimination toward Muslim job applicants. Pers Psychol 63:881–906

Luff S (2010) Why I Blog. Retrieved 18 Jul, from Luff is All You Need. http://luffisallyouneed.blogspot.com/2010/10/why-i-blog.html

McAdam D (1982) Political process and the development of black insurgency, 1930–1970. University of Chicago Press, Chicago, IL

McCarthy JD (1987) Pro-life and pro-choice mobilization: infrastructure deficits and new technologies. In: McCarthy JD, Zald MN (eds) Social movements in an organizational society. Transaction Publishers, New Brunswick, NJ, pp 49–66

Mohanty C (2003) Feminism without borders: decolonizing theory, practicing solidarity. Duke University Press, Durham, NC

Nip JY (2003) The Queer Sisters and its electronic bulletin board: a study of the internet for social movement mobilization. In: Wim van de Donk BD (ed) Cyberprotest: new media, citizens and social movements. Routledge, New York, NY, pp 223–258

Parvez A (2011) Islam and me. Retrieved 19 Jul 2011, from Haute Muslimah. http://www.hautemuslimah.com/2011/05/islam-and-me.html, 9 May

Smith T, Stevens G (1999) The architecture of small networks: strong interaction and dynamic organization in small social systems. Am Sociol Rev 64:403–420

Sobh M (2009) Why is hijab always such a big deal? Retrieved 22 Dec 2011, from HijabTrendz. http://www.hijabtrendz.com/2009/09/22/why-is-hijab-always-such-a-big-deal/, 22 September

Sobh M (2010) The Disney hijab controversy. Retrieved 17 Jul 2011, from Hijabtrendz. http://www.hijabtrendz.com/2010/08/24/disney-hijab-controversy/, 24 August (Also see the comments of the following: Sally, Mariam Sobh, and Kelly, nounou, Natalie, Aisha, aamina, Noor, LK, Zahira, Rania, Samira, Zainab, lilah, and Nadia)

Sobh M (2011a) Super sensitive. Retrieved 22 Dec, from HijabTrendz. http://www.hijabtrendz.com/2011/06/07/super-sensitive/, 7 June (Also see the comments of the following: Carib Muslimah, Azeezah)

Sobh M (2011b) I'm more than a headscarf. Retrieved 17 Jul 2011, from HijabTrendz. http://www.hijabtrendz.com/2011/05/03/headscarf/, 3 May

Sobh M (2011c) Helping American Muslim kids fit in. Retrieved 22 Dec, from HijabTrendz. http://www.hijabtrendz.com/2011/05/09/helping-american-muslim-kids-fit-in/, 9 May

Sobh M (2011d) Top 5 ways to announce to the world you are Mulsim. Retrieved 17 Jul 2011, from HijabTrendz.   http://www.hijabtrendz.com/2011/07/14/top-5-ways-announce-world-muslim/, 14 July

Torres J (2010) Tasteful but trendy. Retrieved 17 Jul, from Recordnet. http://www.recordnet.com/apps/pbcs.dll/article?AID=/20100810/A_LIFE/8100301/-1/A_NEWS07, 10 August

Wilson J (1973) Political organizations. Basics, New York, NY

Editor Biographies

Nitin Agarwal is an associate professor of Information Science at the University of Arkansas at Little Rock. He has a Ph.D. in computer science from Arizona State University with outstanding dissertation recognition. He was recently recognized among the top 20 influential people in their 20s by *Arkansas Business*, a statewide business publication, as creative and talented individuals. He studies the computational aspects of knowledge extraction and prediction from social media and the underlying sociological processes including behavioral modeling, influence, trust, collective wisdom, collective action, crowd dynamics, and community evolution. His research funded by the National Science Foundation (NSF) and the US Office of Naval Research (ONR) leverages fundamentals of data mining, graph mining, content analysis, and large-scale data management. His most recent NSF-funded research focuses on investigating cyber-collective movements from the lens of social media and aims to develop novel socio-computational models to study contemporary forms of collective action. His ONR-funded research examines the role of social media in cultural modeling to help social scientists to quickly analyze both blogs and bloggers at a scale which is otherwise impossible through manual investigation. His research has been extensively published in several leading journals and conferences and includes best paper award and best paper nominees. In 2013, an article he coauthored with Lim and Wigand was awarded the 2012 Best Publication Award in information systems. He has authored two books entitled "Modeling and Data Mining in Blogosphere" published by Morgan & Claypool and "Social Computing in Blogosphere: Challenges, Methodologies, and Opportunities" published by LAP Lambert Academic Publishing AG & Co. KG. He has presented several well-received talks and tutorials on social computing at premier venues, including conferences, academic

institutions, and industry. Videos of these talks are available online at http://
videolectures.net/nitin_agarwal/. To foster an interdisciplinary collaboration
and a synergistic environment, he has edited several books/journal special issues
for IEEE, Elsevier, Springer, and Oxford. He currently serves as program chair and on
the program committees of several prestigious conferences and journals. Dr. Agarwal
has served in the United States National Science Foundation (NSF), United States
Army Research Office (ARO), Canadian Natural Sciences and Engineering Research
Council (NSERC), and Hong Kong's Research Grants Council (RGC) proposal
review panels. One can contact him via http://www.ualr.edu/nxagarwal/.

Merlyna Lim is the Canada Research Chair
in Digital Media and Global Network Society
at Carleton University in Ottawa, Canada.
Currently, she holds a Visiting Research Scholar
position with the Center for Information
Technology at Princeton University. Lim has
published extensively on mutual shaping of
technology and society and political implications
of technology, especially information and
communication technology, in relation to issues
of globalization, power, equality, and democracy.
In her most recent multi-disciplinary research
program, Lim investigates the various roles of digital media in supporting contemporary social movements and transforming politics globally, especially by looking
at the connection of digital media and urban spaces. By reading and analyzing
social movements spatially, Lim's research offers an in-depth understanding of the
relationship between movements, urban space, and digital media. Using empirical
evidence from various contexts (such as the pro-democracy movements in Egypt
and Tunisia, *Bersih* movement in Malaysia, and anticorruption movement in
Indonesia), her research generates conceptual and theoretical frameworks of the
dialectical interplay between digital media and physical urban spaces in the making
of contemporary social movements. Lim holds various awards and fellowships such
as Princeton University CITP Fellowship (2013–2014), KITLV Visiting Fellowship
(Summer 2012), 100 Most Inspiring Indonesian Women from Kartini Foundation
(2011), Our Common Fellowship from the Volkswagen Foundation (2010),
Annenberg Networked Publics Research Fellowship (2005–2006), and Henry
Luce Southeast Asia fellowship (2004). Lim has given more than 150 invited
lectures and presentations, including many keynote addresses, across North
America, the Middle East, Australia, Europe, and Asia. Lim is also an award-
winning author, designer, and visual artist. She won the 2012 Best Publication
Award in Information Systems (2013, with Agarwal and Wigand), the American
Society of Information Technology and Science International Paper Contest (2002),
the Kota Baru Parahyangan Design Competition (with Space for Living Lab, 1999),
and the Autodesk Design-Your-World contest (1997) among others. In 2011, the

Tempe Council of Arts (Arizona, US) selected Lim as one of the featured artists of the year. For more information, see http://www.merlyna.org.

Rolf Wigand is the Maulden-Entergy Chair and Distinguished Professor of Information Science and Business Information Systems at the University of Arkansas at Little Rock. He is the past director of the Graduate Program in Information Management and the founding director of the Center for Digital Commerce both at Syracuse University. His most recent NSF-funded research has focused on the global impact of electronic commerce in ten nations; the impact of electronic commerce on the real estate industry; the software standards development and collective action in the mortgage, automotive parts supply, and RFID in the retail industries; the longitudinal development of trust and leadership in virtual organizations; the analysis of social networks in disaster situations; as well as the tracking and analysis of how sentiments lead to collective action on blogs and social media using novel and innovative methodology. His current NSF research grants (2007–2014) continue with standards development research in the mortgage, retail, and automotive supply industries; the analysis of virtual organizations; networks in disaster settings; as well as the analysis of online blogs and social media as instruments in collective action. He holds or has held editorial positions with numerous journals in the field. His research has appeared in such journals as *MIS Quarterly, Journal of MIS, Sloan Management Review, Journal of Information Technology, Journal of Communication, Electronic Markets: The International Journal, International Journal of Electronic Commerce, International Journal of Information Technology & Decision Making, Knowledge Management: Research & Practice, International Journal of Management, Business & Information Systems Engineering,* and *European Journal of Information Systems.* He is an editorial board and review board member of almost 30 journals, book series, and yearbooks. Wigand is the author of 6 books and over 150 articles, book chapters, and monographs. His book, *Information, Organization and Management: Expanding Markets and Corporate Boundaries* (with A. Picot and R. Reichwald), John Wiley & Sons, was listed among the "75 Best Management Books of All Time" in the *Handelsblatt Management Bibliothek,* Campus Verlag, Frankfurt and New York. Wigand has held visiting professor positions at universities in Munich, Sydney, Helsinki, Mannheim, Stuttgart, Bayreuth, Mexico City, and Tempe/Phoenix, Arizona. He was an invited member to the Joint ACM/AIS Task Force for the Master's Curriculum in Information Systems, i.e., MSIS 2000, MSIS 2006, and MSIS 2013. For more information, see http://www.ualr.edu/rtwigand.